Science Press   Springer-Verlag

Physical Chemistry of Paddy Soils

Edited by

Yu Tian-ren

With 153 Figures and 95 Tables

Science Press, Beijing
Springer-Verlag
Berlin Heidelberg New York Tokyo
1985

Yu Tian-ren
Institute of Soil Science, Academia Sinica
Nanjing
The People's Republic of China

Responsible Editor   Hong Qingwen

Published by Science Press, Beijing

Distribution rights throughout the world, excluding The People's Republic of China, granted to Springer-Verlag Berlin Heidelberg New York Tokyo

ISBN 3-540-13001-2 Springer-Verlag Berlin Heidelberg New York Tokyo
ISBN 0-387-13001-2 Springer-Verlag New York Heidelberg Berlin Tokyo

This work is subject to copyright. All rights are reserved, whether the whole or part of the material is concerned, specifically those of translation, reprinting, re-use of illustrations, broadcasting, reproduction by photocopying machine or similar means, and storage in data banks. Under §54 of the German Copyright Law where copies are made for other than private use, a fee is payable to "Verwertungsgesellschaft Wort" Munich.

© Science Press, Beijing and Springer-Verlag Berlin Heidelberg 1985

Printed in Hong Kong

Printing and binding: C & C Joint Printing Co., (H.K.) Ltd.
2131/3140-54321

Science Press Book No. 4246-53

PREFACE

Soil physical chemistry is a science dealing with the interrelationship between physical and chemical phenomena in soils. The physico-chemical properties of paddy soil follow the general behavior of soils, and also have their own characteristics. Studies on the physical chemistry of paddy soil should proceed from this viewpoint.

Research work on this subject in the Department of Soil Electrochemistry, Institute of Soil Science, began in the early fifties. Our starting point was that most of the physico-chemical characteristics of paddy soil should be related to "water", the first word of the term "paddy soil" in Chinese. Among these characteristics the periodical change in the oxidation-reduction process appears to be the most distinct and important. Therefore, this subject was inevitably the core of our studies. We were also aware that the exploitation of research fields could not be separated from the development of new research methods. In the last thirty years our researches have principally been based on these considerations.

This book is a monograph summarizing our researches in this field. Except for the limited space necessary for the explanation of basic principles, discussions are based almost exclusively on experimental results obtained in China. As the scope of physico-chemical properties of paddy soil is rather wide, the discussions here are concentrated on the most active and at the same time the most essential areas, namely interactions among charged particles (electrons, protons, ions and clay) and their chemical consequences. Starting from the characteristic features of paddy soil, special attention is paid to the transfer of electrons (oxidation-reduction reaction), with half of the contents (five chapters) devoted to this subject. First, the intensity factor (redox potential) and the capacity factor (quantity of reducing substances) of the oxidation-reduction property are dealt with, and then separate discussions on several important redox systems (oxygen, iron, manganese and sulfur) follow successively. The second part of the book, including ion adsorption, acidity and electrical conductivity, deals mainly with the interaction between soil clay and ions or protons. The reader can see that most of these properties are also related directly or indirectly to those included in the first part of the book. The third part of the book deals with the applied aspect of the physical chemistry of paddy soil, namely its relations with soil genesis and plant growth. In discussing the latter question we not only regard the soil as a medium in which plants are grown and by which they are influenced, but also consider the influence of the plants on the properties of the soil.

We are aware that as a monograph both the scope and the depth of this book are far from satisfactory. We hope, however, that the materials presented may be of some use and interest to the study of this field. There are many theoretical questions concerning the physico–chemical properties of paddy soil, such as the nature of organic redox systems, the quantitative relationship between redox potential and various redox systems, the interrelationship between electron activity (pe) and proton activity (pH), the chemical mechanism of pH change during the alternation of oxidation condition and reduction condition, and the real cause of poor plant growth under different reducing conditions, remaining to be further elucidated.

This book was originally written in Chinese. Inasmuch as almost all of the research work conducted in China in this respect is virtually unknown in foreign countries, we consider it appropriate to render it into English. It is our hope that the publication of this monograph in English will be helpful to the mutual understanding between western and Chinese soil scientists, and to further progress in the physical chemistry of soils, especially of paddy soils.

The materials cited in this book, unless referring to the literature listed at the end of each chapter, are unpublished data of the respective author(s).

YU TIAN–REN

Head, Department of Soil Electro-chemistry,
Institute of Soil Science, Academia Sinica

CONTENTS

Preface v

CHAPTER 1 OXIDATION–REDUCTION POTENTIAL

1.1 Oxidation–reduction potential and electron activity and proton activity 1
    1.1.1 Eh and pe 1
    1.1.2 Quantitative relationship between Eh and electron donor or electron acceptor 3
    1.1.3 Effect of proton activity (pH) on Eh 4
1.2 Eh in relation to chemical equilibria among oxidation–reduction systems in paddy soils 6
    1.2.1 Sequential reduction 6
    1.2.2 Platinum electrode and problems on the measurement of Eh 9
    1.2.3 Eh-determining systems 13
    1.2.4 Poising 15
    1.2.5 Gradation of oxidation–reduction status according to Eh 16
1.3 Heterogeneity in oxidation–reduction potential in paddy soils 18
    1.3.1 Oxidizing layer and reducing layer near water–soil interface 18
    1.3.2 Inner and outer parts of soil clods and cleavage surface 19
    1.3.3 Centrifugate, suspension and bulk soil 20
    1.3.4 Root–zone and bulk soil 21
1.4 Oxidation–reduction potential of paddy soils 21
    1.4.1 Oxidation–reduction potential of paddy soils with different water regimes .. 21
    1.4.2 Dynamics of oxidation–reduction potential in the profile 22
    1.4.3 Oxidation–reduction potential of representative paddy soils 25

CHAPTER 2 REDUCING SUBSTANCES

2.1 Characterization of reducing substances of the soil 27
    2.1.1 Reducing intensity 27
    2.1.2 Chemical activity 29
    2.1.3 Kinds and properties 30
2.2 Factors affecting the amount of reducing substances in paddy soils 33
    2.2.1 Organic matter 33
    2.2.2 Eh 34
    2.2.3 pH 36
2.3 Dynamics of reducing substances in paddy soils 37
    2.3.1 Quantity 37
    2.3.2 Changes in organic and inorganic components 38
    2.3.3 Changes in fractions with different reduction intensity 41
2.4 Amount of reducing substances in paddy soils 42
    2.4.1 Heterogeneity in distribution of reducing substances in the soil 42
    2.4.2 Distribution in the profile 43
    2.4.3 Amount in paddy soils with different oxidation–reduction regime ... 44

CHAPTER 3 OXYGEN

3.1 Characteristics of the oxygen system 47

      3.1.1 Oxygen as an oxidation–reduction system 47
      3.1.2 Relationship between oxygen concentration and *Eh* 48
      3.1.3 Voltammetric behavior of oxygen 50
      3.1.4 Diffusion of oxygen 52
  3.2 Source and consumption of oxygen in paddy soils 53
      3.2.1 Source of oxygen 53
      3.2.2 Consumption of oxygen 54
      3.2.3 Dynamic balance of oxygen 58
  3.3 Oxygen content of paddy soils 62
      3.3.1 Heterogeneity in oxygen content of paddy soils 62
      3.3.2 Distribution in the surface layer 63
      3.3.3 Distribution in the profile 66
      3.3.4 Effect of water management 67

CHAPTER 4 IRON AND MANGANESE

  4.1 Chemical behavior of iron and manganese in soil 69
      4.1.1 Effect of *pe* and pH 69
      4.1.2 Effect of O_2 and CO_2 partial pressure 69
      4.1.3 Sensitivity to oxidation–reduction condition and acidity 70
  4.2 Iron in paddy soils 73
      4.2.1 Amount of active iron 73
      4.2.2 Forms of ferrous iron 74
      4.2.3 Physico-chemical equilibria among various forms of ferrous iron.... 82
  4.3 Manganese in paddy soils 83
      4.3.1 Amount of active manganese 84
      4.3.2 Relationship between amount of exchangeable manganese and pH.... 84
      4.3.3 Chelation with organic substances 86
  4.4 Dynamics of iron and manganese in paddy soils 88
      4.4.1 Total amounts of reduced iron and manganese 88
      4.4.2 Water-soluble ferrous iron 89
      4.4.3 Changes in the profile 90

CHAPTER 5 SULFUR

  5.1 Reduction of sulfate in soils 92
      5.1.1 Effect of *Eh* and pH 92
      5.1.2 Effect of organic matter 93
      5.1.3 Effect of temperature 95
  5.2 Factors affecting the chemical equilibria of sulfides in soils 95
      5.2.1 pH 95
      5.2.2 Fe^{2+} 98
      5.2.3 Mn^{2+} and Zn^{2+} 102
      5.2.4 Amount of sulfide 104
  5.3 Dynamics of sulfides in paddy soils 105
      5.3.1 Acid sulfate soil 105
      5.3.2 Loamy paddy soil 106
      5.3.3 Paddy soil derived from red soil 107
  5.4 Forms and amount of sulfur in paddy soils 108

CHAPTER 6 ION ADSORPTION

  6.1 Electric charge of paddy soils 111
      6.1.1 Property of electric charge 111
      6.1.2 Quantity of electric charge 112
      6.1.3 Surface charge density 115

6.2 Interactions between ions and soil particles 116
    6.2.1 Ion adsorption in relation to electric charge of the soil 116
    6.2.2 Bonding strength of cations with the soil 117
    6.2.3 Dissociation of adsorbed cations 122
 6.3 Composition of exchangeable cations in paddy soils 123
    6.3.1 Characteristics in the composition of exchangeable cations in paddy soil.. 124
    6.3.2 Composition of exchangeable bases in different types of paddy soil.... 125
 6.4 Base-saturation percentage in paddy soils 126
    6.4.1 Base-saturation percentage in relation to pH 126
    6.4.2 Base-saturation percentage in relation to individual cation-saturation percentage 127
    6.4.3 Influence of soil management measures on base-saturation percentage.... 128
    6.4.4 Base-saturation percentage of principal types of paddy soil........... 129

CHAPTER 7 ACIDITY

 7.1 Indexes of soil acidity 131
    7.1.1 pH 131
    7.1.2 Lime potential 134
    7.1.3 pK 137
    7.1.4 Exchange acidity and exchange alkalinity 139
 7.2 Characteristics in the acidity of paddy soil............................. 141
    7.2.1 Change in acidity during submergence 141
    7.2.2 Comparison between paddy soil and its parent soil 146
    7.2.3 Peculiarity of acid sulfate soil 150
 7.3 Buffering capacity of paddy soils 151
    7.3.1 Buffering capacity in different horizons 151
    7.3.2 Buffering capacity of different types of soil 153
    7.3.3 Buffering capacity in relation to soil fertility 154
 7.4 Acidity regime of principal types of paddy soil 155

CHAPTER 8 ELECTRICAL CONDUCTIVITY

 8.1 Factors affecting the electrical conductivity of soils 157
    8.1.1 Interactions between clay and ions 157
    8.1.2 Physical factors 159
    8.1.3 Soluble salts 162
 8.2 Characteristics of electrical conductivity of paddy soil................ 164
    8.2.1 Comparison between paddy soil and other types of soil............ 164
    8.2.2 Change in electrical conductivity during submergence 165
 8.3 Electrical conductivity in relation to the fertility status of paddy soil 169
    8.3.1 Relationship between electrical conductivity and fertility level of the soil .. 169
    8.3.2 Effect of agricultural measures on electrical conductivity 170
    8.3.3 Change in electrical conductivity due to the uptake of ions by rice 172
    8.3.4 Electrical conductivity and desalinization of salt-affected soils 173
 8.4 Electrical conductivity of principal types of paddy soil 174
    8.4.1 Weakly-leached paddy soil 174
    8.4.2 Moderately-leached paddy soil 174
    8.4.3 Strongly-leached paddy soil 177

CHAPTER 9 PHYSICAL CHEMISTRY OF PADDY SOIL IN RELATION TO SOIL GENESIS

 9.1 Genesis of paddy soil 178

    9.1.1 Solution 178
    9.1.2 Reduction... 181
    9.1.3 Complexation 184
  9.2 Change in clay minerals during the genesis of paddy soil................. 186
    9.2.1 Depotassication 186
    9.2.2 Deferrugination 188
    9.2.3 Other changes 189
  9.3 Classification and types of paddy soil 192
    9.3.1 Principles of classification 192
    9.3.2 Weakly–leached paddy soil 193
    9.3.3 Moderately–leached paddy soil 194
    9.3.4 Strongly–leached paddy soil 194

CHAPTER 10   SOIL AND PLANTS

10.1 Oxidation–reduction status 197
    10.1.1 Adaptation of rice plants to oxidation–reduction conditions of the soil .. 197
    10.1.2 Influence of rice root on the oxidation–reduction status of the soil.... 197
    10.1.3 Influence of the oxidation–reduction condition of the soil on the Eh of plants 201
    10.1.4 Toxicity problems under strong reducing conditions 203
10.2 pH 205
10.3 Physico–chemical characteristics of paddy soil in relation to nutrient supply.. 206
    10.3.1 Forms of nutrients 207
    10.3.2 Movement of nutrients 208
10.4 Nutrient content of the soil in relation to uptake by rice................... 214

CHAPTER 1

OXIDATION–REDUCTION POTENTIAL

Liu Zhi-guang

The most distinct physico–chemical change in paddy soils is the periodical change of the oxidation–reduction regime. Oxidation–reduction potential (Eh) is the most important index for characterizing the degree of oxidation or reduction of a soil. It reflects the equilibrium position among various redox systems (oxygen, iron, manganese, nitrogen, sulfur, carbon) when they exist in steady equilibrium. and determines the reacting direction of these systems when the equilibrium has not been attained. The oxidation–reduction potential of paddy soils is characterized by the wide range of variation, i.e., from about minus three hundred millivolts to plus seven hundred millivolts, nearly covering the whole variation range commonly encountered in the biosphere. This wide-range variation of oxidation–reduction potential causes many chemical properties of the soil to change distinctly, and exerts profound influence on the growth of plants in many ways.

In this chapter the relationships between oxidation–reduction potential and electron activity and proton activity are briefly discussed. Special reference is made to the interrelations between oxidation–reduction potential and the equilibria among various redox systems and, finally, to the oxidation–reduction potential under different conditions in connection with the heterogeneity in the oxidation–reduction regime of paddy soils.

1.1 OXIDATION–REDUCTION POTENTIAL AND ELECTRON ACTIVITY AND PROTON ACTIVITY

1.1.1 *Eh* and *pe*

Electron activity (a_e) is an index for expressing the quantity of electrons in a system. Like the definition for the activity of other substances, it denotes the contribution of electrons to the partial molar free energy of the system. It can be expressed as the negative logarithm of the activity of electrons (pe), just as pH for the activity of protons:

$$pe = -\log a_e \tag{1-1}$$

For the oxidation–reduction reaction

$$(\text{oxidant}) + e \rightleftharpoons (\text{reductant}) \tag{1-2}$$

at equilibrium

$$K = \frac{(\text{reductant})}{(\text{oxidant})(e)} \tag{1-3}$$

under standard conditions

$$\log K = pe^0 \tag{1-4}$$

where K is the equilibrium constant. The equilibrium constant is related to the change in standard free energy (ΔG^0) of a reaction as follows:

$$\Delta G^0 = -RT\ln K = -nFE^0 \tag{1-5}$$

hence

$$E^0 = \frac{RT}{nF}\ln K \tag{1-6}$$

If $n=1$, at 25°C

$$E^0 = 0.059\ \log K = 0.059 pe^0 \tag{1-6a}$$

In Equation (1-5) the numeral value of ΔG^0 is equal to the difference between the sum of the change in standard free energy of formation of the products and that of the reactants in a reaction, i.e.,

$$\Delta G^0 = \Sigma \Delta G^0_{f(\text{product})} - \Sigma \Delta G^0_{f(\text{reactant})}$$

Considering the role of the electron in the reduction of oxygen:

$$O_2 + 4H^+ + 4e^- \rightleftharpoons 2H_2O \tag{1-7}$$

$$K = \frac{1}{(O_2)(H^+)^4(e)^4} \tag{1-7a}$$

substituting the numeral value of $10^{20.8}$ for K in Equation (1-7a) and taking logarithmic form, we get:

$$pe = 20.8 + 1/4\ \log P_{O_2} - pH \tag{1-8}$$

Numeral values of pe^0 and E^0 for some oxidation–reduction systems in submerged soils are tabulated in Table 1-1.

TABLE 1-1 pe^0 and E^0 of some redox systems in submerged soils (25°C)[10]

System	$pe^0\ (=\log K)$	E^0 (V)
$\frac{1}{4}O_2 + H^+ + e \rightleftharpoons \frac{1}{2}H_2O$	20.8	1.23
$\frac{1}{2}MnO_2 + 2H^+ + e \rightleftharpoons \frac{1}{2}Mn^{2+} + H_2O$	20.8	1.23
$Fe(OH)_3 + 3H^+ + e \rightleftharpoons Fe^{2+} + 3H_2O$	17.9	1.06
$\frac{1}{2}NO_3^- + H^+ + e \rightleftharpoons \frac{1}{2}NO_2^- + \frac{1}{2}H_2O$	14.1	0.83
$\frac{1}{8}SO_4^{2-} + \frac{5}{4}H^+ + e \rightleftharpoons \frac{1}{8}H_2S + \frac{1}{2}H_2O$	5.12	0.30
$\frac{1}{8}CO_2 + H^+ + e \rightleftharpoons \frac{1}{8}CH_4 + \frac{1}{4}H_2O$	2.86	0.17
$H^+ + e \rightleftharpoons \frac{1}{2}H_2$	0	0

The energy gained in the transfer of one mole of electrons to an oxidant from H_2, expressed in volts, is the oxidation–reduction potential (Eh). Eh is related to pe as follows (when $n=1$):

$$pe = \frac{F}{2.303RT}Eh \tag{1-9}$$

1.1 Oxidation-reduction potential and electron activity and proton activity

at 25°C

$$pe = \frac{Eh}{0.059} \qquad (1-10)$$

Because the Eh of a system can be directly measured, it is a commonly used index.

Fig. 1-1 shows the relationship between Eh and pe in oxidation–reduction equilibria of some natural redox systems (including soil). In the figure, the uppermost straight line represents the equilibrium between O_2 at one atmosphere and water, and the lowermost straight line that of H_2 at one atmosphere in equilibrium with water. The region within the circle stands for the actually measured range for paddy soils.

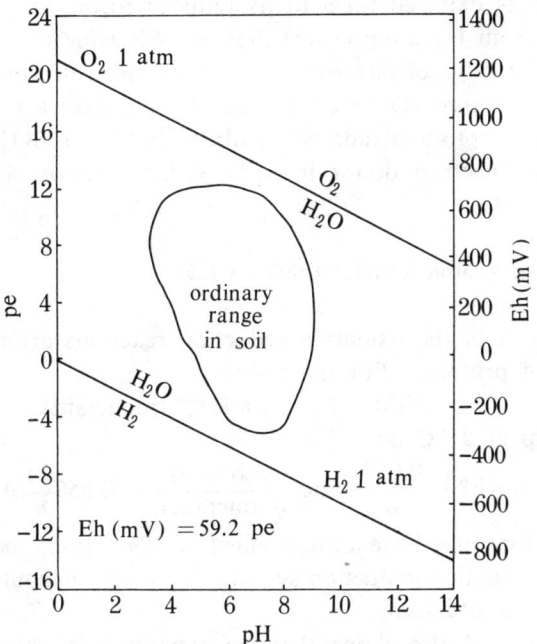

Fig. 1-1 Relationship between Eh and pe in oxidation–reduction equilibrium of redox systems in water (including soil) (25°C)

The pe of paddy soils generally fluctuates within the range of 10 to -5 at pH 7. The high value of pe represents an oxidizing condition where electron activity is low and oxidation–reduction potential is high, whereas a low or even negative value of pe denotes a reducing condition.

1.1.2 Quantitative relationship between Eh and electron donor or electron acceptor

In an oxidation–reduction reaction, oxidants are reduced through accepting electrons from other substances, and reductants are oxidized through donating electrons to other substances. Therefore, oxidants are electron acceptors, and reductants electron donors.

For the generalized oxidation–reduction reaction:

$$(\text{oxidant}) + ne \rightleftarrows (\text{reductant})$$

at equilibrium there is a quantitative relationship between Eh and electron donor (reductant) and electron acceptor (oxidant) as follows:

$$Eh = E^0 + \frac{RT}{nF} \ln \frac{(\text{electron acceptor})}{(\text{electron donor})} \quad (1\text{-}11)$$

at 25°C

$$Eh = E^0 + \frac{0.059}{n} \log \frac{(\text{electron acceptor})}{(\text{electron donor})} \quad (1\text{-}12)$$

In the equations E^0 is the standard potential, representing the ability of an oxidation–reduction system (couple) to donate or accept electrons under standard conditions. For a given system the ratio of electron acceptor to electron donor is the ratio of its oxidized form to its reduced form.

It is seen from Equation (1-11) that the Eh value of a system is related to both E^0 and the ratio of electron acceptor to electron donor. In paddy soils oxygen is the most important oxidant and also the most active electron acceptor. Other electron acceptors include NO_3^-, Mn^{4+}, Fe^{3+} and SO_4^{2-}. Organic matter is the principal electron donor in soil. Other electron donors include Fe^{2+}, Mn^{2+}, S^{2-} and H_2.

1.1.3 Effect of proton activity (pH) on Eh

In soils most of the oxidation–reduction reactions are accompanied by the participation of protons. For the reaction:

$$(\text{oxidant}) + ne + mH^+ \rightleftarrows (\text{reductant})$$

the relationship at 25°C is:

$$Eh = E^0 + \frac{0.059}{n} \log \frac{(\text{oxidant})}{(\text{reductant})} - 0.059 \frac{m}{n} \text{pH} \quad (1\text{-}13)$$

Thus, H^+ ions may have a direct effect on Eh. It can be seen in Table 1-1 that all of the oxidation–reduction systems commonly encountered in paddy soils belong to such a situation.

Equation (1-13) also shows that the magnitude of the effect of pH on Eh is dependent on the ratio m/n of the reaction. When the ratio m/n or H^+/e is equal to 1:

$$Eh = E^0 + \frac{0.059}{n} \log \frac{(\text{oxidant})}{(\text{reductant})} - 0.059 \text{pH} \quad (1\text{-}14)$$

This means that the theoretical value of the quotient $\Delta Eh / \Delta \text{pH}$ is -59 mV at 25°C. This is just what happens in the oxygen–H_2O system and some organic systems, such as quinone–hydroquinone.

For the O_2–H_2O system:

$$Eh = 1.23 + 0.015 \log P_{O_2} - 0.059 \text{pH} \quad (1\text{-}15)$$

For the quinone–hydroquinone system at a pH lower than about 8:

$$Eh = 0.699 + 0.0295 \log \frac{(\text{quinone})}{(\text{hydroquinone})} - 0.059 \text{pH} \quad (1\text{-}16)$$

In some oxidation–reduction reactions m/n is greater than 1. For example,

1.1 Oxidation-reduction potential and electron activity and proton activity

it can be seen from Table 1-1 that for the $Fe(OH)_3 - Fe^{2+}$ system the quotient m/n is $3/1=3$. Hence:

$$Eh = 1.06 - 0.059 \log Fe^{2+} - 0.177 pH \quad (1-17)$$

Actually, in paddy soils there is a variety of oxidation-reduction systems. The interactions of these systems with H^+ ions are also quite complicated. Hence the quotient m/n varies considerably, and so does the quotient $\Delta Eh/\Delta pH$. Experiments showed that under natural conditions the quotient $\Delta Eh/\Delta pH$ was affected by the ratio of the amount of organic reducing substances to that of ferrous iron. For instance, Fig. 1-2 shows that the quotient for the water-

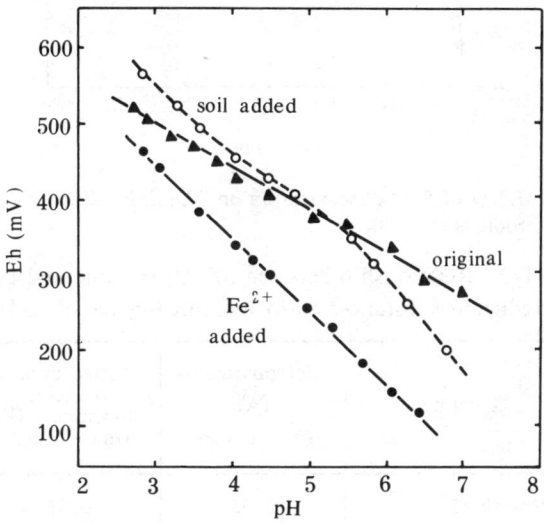

Fig. 1-2 Eh-pH curves of water-soluble decomposition products of milk vetch

soluble decomposition products of milk vetch increases from the original value of 53 mV to 91 mV after the addition of some paddy soil and to 102 mV after the addition of 50 ppm of Fe^{2+}. This means that the quotient is larger when the effect of Fe^{2+} is great, and smaller when the effect of organic reducing substances is great. It is for the same reason that the higher the concentration of Fe^{2+}, the larger the quotient for both the original water-soluble decomposition products of milk vetch and its diluted solution (Fig. 1-3). Measurements for the centrifugates of incubated soils also indicated that the higher the proportion of ferrous iron, the larger the quotient (Table 1-2). This seems to imply that for paddy soils with various oxidation-reduction systems the numeral value of $\Delta Eh/\Delta pH$ is determined by the kinds and quantities of relevant systems. As a consequence, the measured $\Delta Eh/\Delta pH$ varies greatly for different paddy soils and for different horizons of the same soil (Table 1-3). On the other hand, the quotient for red soils, yellow soils and some other upland soils approximates to the theoretical value of oxygen, namely 60 mV at 30°C[5].

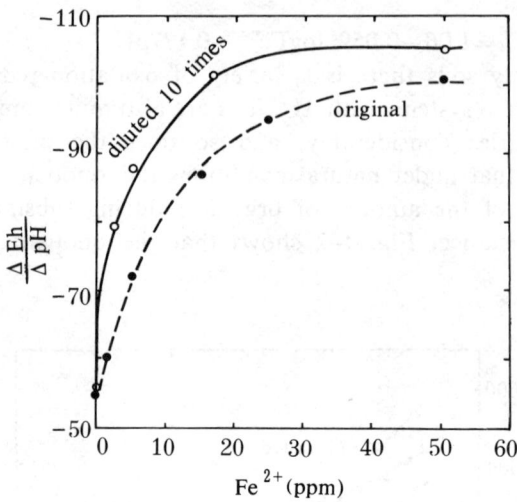

Fig. 1-3 Effect of Fe^{2+} concentration on $\Delta Eh/\Delta pH$ value for the decomposition products of milk vetch

TABLE 1-2 Relationship between $\Delta Eh/\Delta pH$ value and Fe^{2+}/active organic reducing substances ratio in centrifugates of soils

Soil	Treatment	Ferrous iron (A) (m.e. / 100g)	Active organic reducing substances (B) (m.e. / 100g)	$\dfrac{(A)}{(B)}$	$\dfrac{\Delta Eh}{\Delta pH}$
Paddy from red soil	Radish (5%)	3.35	0.81	4.13	104
	Vetch (5%)	1.31	0.93	1.41	104
	Milk vetch (5%)	1.76	2.32	0.76	102
	Milk vetch (3%)	0.77	1.75	0.44	93
	Radish (3%)	0.19	1.22	0.16	83
Acid sulfate	—	0.09	0.08	1.13	101
Gleyed	—	0.18	0.28	0.64	93
Loamy	—	0.014	0.11	0.13	80
Sandy	Milk vetch (3%)	0.021	1.27	0.017	59

1.2 Eh IN RELATION TO CHEMICAL EQUILIBRIA AMONG OXIDATION–REDUCTION SYSTEMS IN PADDY SOILS

1.2.1 Sequential reduction

In Table 1-4 are tabulated in descending order the theoretical E^0 values

1.2 Eh in relation to chemical equilibria in paddy soils

TABLE 1-3 Relationship between Eh and pH for different horizons of paddy soils after submerging[5]

Locality	Horizon	Eh_6 (mV)	$\dfrac{\Delta Eh}{\Delta pH}$
Ganzhou	A (surface)	280	90
	P (plowpan)	278	40
	B (illuvial)	480	38
	G (gley)	487	35
Dongxiang	A (surface)	212	110
	P (plowpan)	300	100
	B (illuvial)	458	65
	G (gley)	523	50
Qingjiang	A (surface)	260	130
	P (plowpan)	463	75
	B (illuvial)	493	53
	G (gley)	505	35

and practically measured Eh values of some oxidation–reduction systems commonly encountered in paddy soils. Since the pH of most natural systems including paddy soil is ordinarily around 7, all the Eh are corrected to this pH.

TABLE 1-4 Sequential reduction of some electron acceptors in paddy soils[2]

Redox reaction	E^0 at pH 7 (V)	Actually measured Eh (V)
$O_2 + 4H^+ + 4e \rightleftarrows 2H_2O$	0.82	0.65—0.3
$NO_3^- + 2H^+ + 2e \rightleftarrows NO_2^- + H_2O$	0.54	0.5—0.2
$MnO_2 + 4H^+ + 2e \rightleftarrows Mn^{2+} + 2H_2O$	0.43	0.4—0.2
$FeOOH + 3H^+ + e \rightleftarrows Fe^{2+} + 2H_2O$	0.17	0.3—0.1
Organic systems	−0.03—(−0.2)	0—(−0.2)
$SO_4^{2-} + 9H^+ + 6e \rightleftarrows HS^- + 4H_2O$	−0.16	0—(−0.15)
$2H^+ + 2e \rightleftarrows H_2$	−0.414	−0.15—(−0.3)
Methane fermentation	—	−0.2—(−0.3)

In paddy soils, the oxidation–reduction reaction among various redox systems generally proceeds following the order listed in Table 1–4. An upper system in the table with a stronger affinity for electrons can oxidize a lower system. The larger the difference in E^0 between two systems, the more easily the oxidation–reduction reaction will proceed.

The practically measured Eh range for a system represents the Eh range within which the system can participate in an oxidation–reduction reaction. Theoretically, starting from the upper limit of the Eh range of a given system, the oxidized form of that system begins to be reduced; and when the Eh falls to the lower limit of the Eh range where its oxidized form has been wholly transformed to its reduced form, the oxidized form of the next system with a lower E^0 begins to be reduced. For instance, only after the complete reduction of FeOOH can SO_4^{2-} be reduced with the formation of S^{2-}, and only after the complete disappearance of SO_4^{2-} can there be the formation of H_2 and CH_4. However, because of the participation of microorganisms and the difference in reaction velocities of various systems, it can be found that two simultaneously proceeding reactions in a soil may overlap, especially when the E^0 of two systems do not differ greatly. This is the basic reason why the Eh ranges of several systems may overlap, as is shown in Table 1–4.

It is for the same reason that the lowering of the Eh after the submerging of a soil can be retarded by the addition of a substance having a strong power of accepting electrons. And, if the amount added is sufficiently large, the Eh of the soil may even be maintained within a certain range for a certain period. Such an example is shown in Fig. 1–4. It can be seen that the effect of O_2 can last about two days, and that of NO_3^- twelve days. Similarly, the Eh of a reduced paddy soil increases immediately after the addition of KNO_3 or O_2,

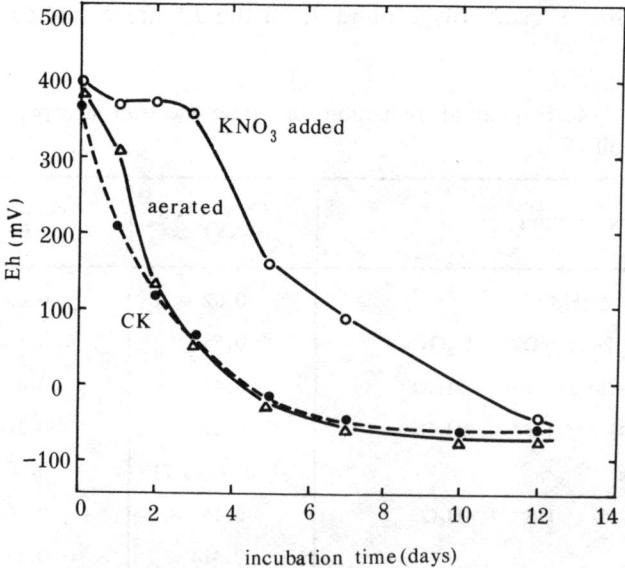

Fig. 1–4 Retardation of Eh-lowering after submerging by oxidizing agents (paddy soil derived from red soil)

1.2 Eh in relation to chemical equilibria in paddy soils

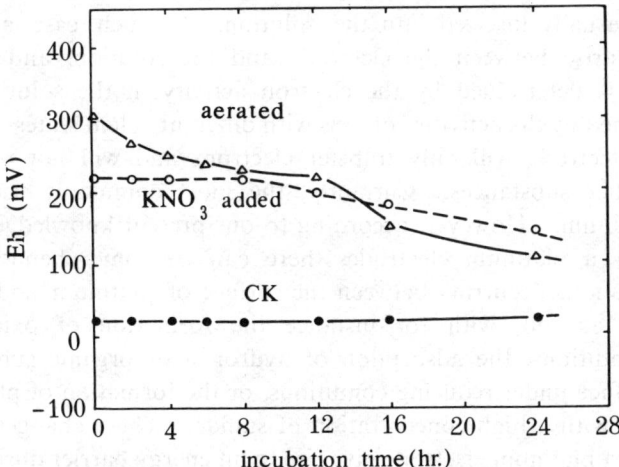

Fig. 1-5 Effect of oxidizing agents on Eh of reduced soil (paddy soil derived from red soil)

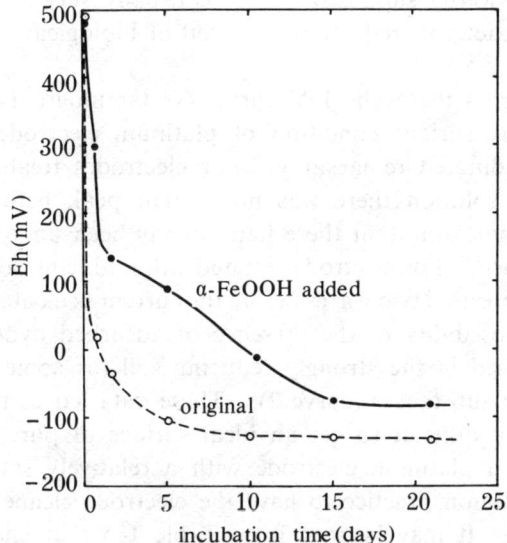

Fig. 1-6 Retardation of Eh-lowering after submerging by α-FeOOH (sandy paddy soil)

and it can be maintained within a certain range for one day (Fig. 1-5). Even α-FeOOH with an oxidizing power that is not so strong can exert a retarding effect on the lowering of the Eh during the submergence of the soil (Fig. 1-6).

1.2.2 Platinum electrode and problems on the measurement of Eh

In the above sections mention has been made of the significance of the measurement of oxidation–reduction potential. For its measurement an inert metal electrode is usually inserted into the solution. In such case a potential difference will arise between the electrode and the solution, and the magnitude of potential is determined by the electron activity in the solution which is in turn determined by the activities of ions with different valent states. Theoretically, an "inert" electrode will only transfer electrons and will not react chemically with any other substances. Currently, the most commonly used electrode is made of platinum. However, according to our present knowledge of the surface properties of a platinum electrode, there can be some chemical or physico–chemical reactions occurring between the surface of platinum and the measured system such as soil, with for instance the formation of oxide films under oxidizing conditions, the adsorption of hydrogen or organic substances on the electrode surface under reducing conditions, or the formation of platinum sulfides in a solution with a high concentration of sulfide. These changes in the surface properties of a platinum electrode give rise to an energy barrier during the transfer of electrons at the electrode surface and, as a result, the transfer rates of electrons on platinum electrodes with different surface properties may differ considerably. Therefore, the platinum electrode does not function as a perfectly reversible electrode as was previously supposed, and it is usually very difficult to precisely measure the actual oxidation–reduction potential of biological systems, including that of soil.

It can be seen from the cyclic I–V curves (voltammograms) shown in Figs. 1–7 and 1–8 that the surface condition of platinum electrodes after different chemical treatments differed remarkably. For electrodes treated with H_2O_2 or dichromate cleaning solution there was no current peak before 0.5 V (curves 1 and 2, Fig. 1–7), indicating that there had already been an oxide film present on the electrode surface. For electrodes treated with reducing solution to remove the oxide film there appeared two waves before the current peak at 0.91 V, implying that there was the possibility of the presence of adsorbed hydrogen etc. If the electrode was immersed in the strongly reducing soil for some time (Fig. 1–8), it could adsorb other substances (curve 2). These data led us to the conclusion that it would be very difficult to get an ideal surface of pure platinum metal.

In order to get a platinum electrode with a relatively satisfactory surface condition, it is a common practice to have the electrode cleaned or "activated" in a variety of ways. It may be seen from Table 1–5 that chemical treatment of the electrode with coating–removing solution helps accelerate the establishment of equilibrium potential and improves the reproducibility of results.

When measurements are made with the platinum electrode in paddy soils with several oxidation–reduction systems, there is the problem of mixed potential[4]. This mixed potential differs from the equilibrium potential of a mixed system in that it is a weight average of the respective potential of each system. The contribution of each system to electrode potential is determined by the electron exchange current between that system and the electrode, and the current is in turn dependent on the rate constant of the chemical reaction of the oxidation–reduction system *per se* and that of electron exchange between this system and the electrode. The surface condition of the platinum electrode may affect the

1.2 *Eh* in relation to chemical equilibria in paddy soils

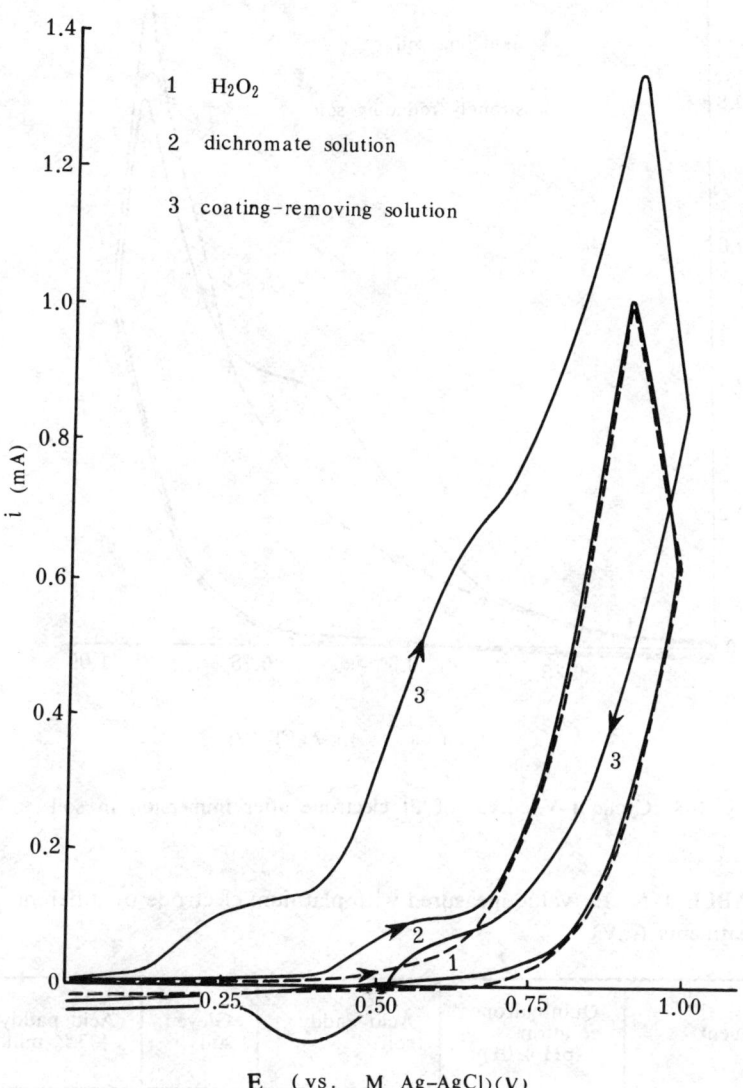

Fig. 1-7 Cyclic I-V curves of Pt electrode after chemical treatments

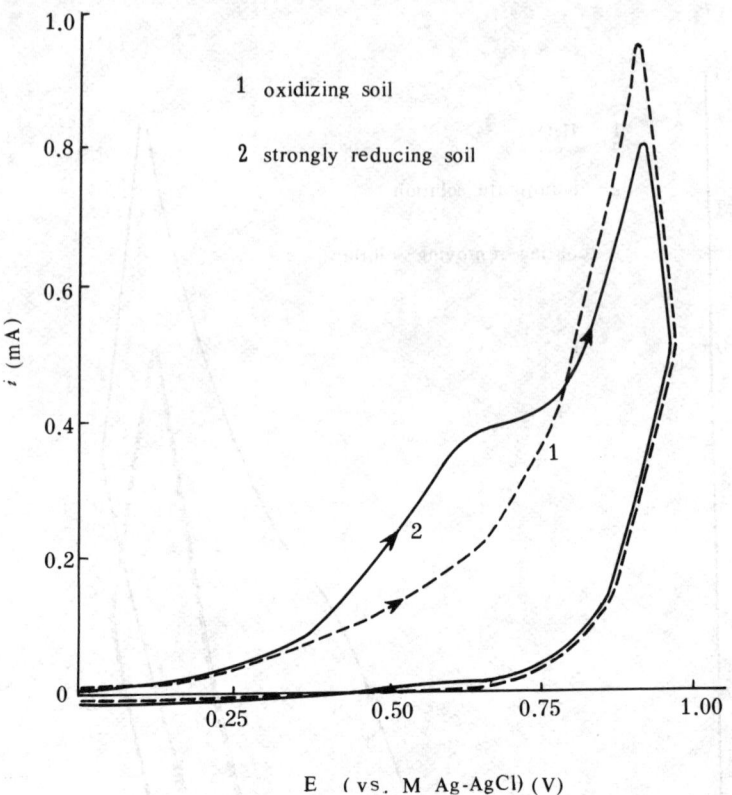

Fig. 1-8 Cyclic I-V curves of Pt electrode after immersion in soil

TABLE 1-5 Eh value measured with platinum electrode by different treatments (mV)

Treatment	Quinhydrone solution (pH 4.01)	Acid paddy soil	Gleyed soil	Acid paddy soil + 3% milk vetch
Film–removing solution	218	388	175	48
Detergent and H_2O_2	232	420	202	68
Conc. HNO_3 and NH_4OH	225	411	194	57
Ignition after immersion in cleaning solution	221	394	182	51
Equilibrium value	218	385	170	45

1.2 Eh in relation to chemical equilibria in paddy soils

measured mixed potential through affecting the electron exchange current.

If the relative concentration of the oxidized form or the reduced form of an oxidation–reduction system is very low, as for instance when the ratio of the two forms is greater than 10^3 or smaller than 10^{-3}, or if the absolute concentration of the system is low, the electron exchange current contributed by the given substance may be so small that it is not sufficient to cause the electrode to correctly register the oxidation–reduction potential of that substance. In such cases other substances with relatively larger electron exchange currents will play a greater or dominant role. If the concentrations of all the oxidation–reduction systems are low, the potential reading will be very unstable, and the reproducibility poor.

It is due to the reasons mentioned above that it is generally very difficult for a platinum electrode to register the actual Eh of the oxidation–reduction system of the soil within a short time. The commonly measured oxidation–reduction potentials, including many of the data cited in this book, are approximate values which are affected by a variety of measuring conditions. In order to get an accurate result, it is necessary to treat the electrode surface carefully and let the electrode be in contact with the soil for several hours.

Since the electrode potential is a function of time after contact of the electrode with the medium, it is possible to use an alternate method for the measurement of the Eh of the soil within a short time by extrapolation. The principle of the method is briefly described as follows: Let the electrode potential deviate from the equilibrium potential positively or negatively by polarization of the electrode. Then, monitor the change in electrode potential with time caused by depolarization due to reducing or oxidizing substances of the soil. The two depolarization curves which are usually straight lines are extrapolated, and the intersection is taken as the equilibrium potential of the soil. Fig. 1–9 contains some examples. According to results from several tens of paddy soils, the values of oxidation–reduction potential measured in this way deviated from the equilibrium potential by no more than 10 mV in most cases[9].

1.2.3 Eh-determining systems

The principal oxidation–reduction systems in paddy soils are oxygen, organic substances, iron and manganese. The contribution of each system to electrode potential is determined by the electron exchange current density between that system and the electrode, and the current density is in turn determined by the concentration of that system and the rate constant at the electrode. Under conditions in which the difference in rate constant is not so large, as is usually the case, the relative concentration of each system will be the chief factor, and the system which is highest in concentration will be dominant in controlling the electrode potential.

Based on observations on a well-drained "White soil", the Eh-determining system under different conditions may be imagined as follows. Before submergence, the pores of the cultivated layer are filled with air, and except for acid soils there is little possibility for the existence of ferrous iron or manganous ions.

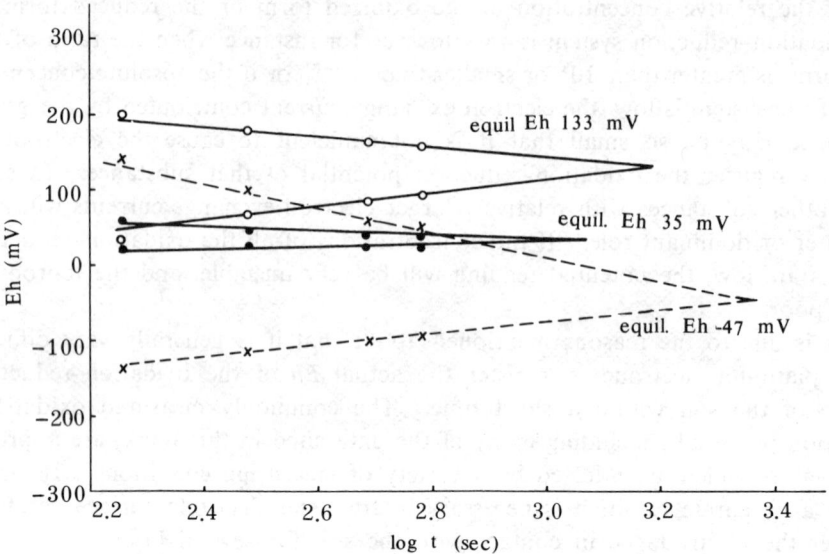

Fig. 1-9 Measurement of Eh of the soil by extrapolation of depolarization curves[9]

Under such circumstances oxygen plays a dominant role in determining the oxidation-reduction potential of the soil due to its high concentration, and causes the Eh to be higher than 300 mV, although the O_2-H_2O system is not reversible with respect to the electrochemical reaction at the platinum electrode. After submergence, the source of oxygen from the atmosphere is cut off, and organic reducing substances are produced gradually at the expense of oxygen due to the biological activities of micro-organisms. Under such conditions organic systems will play an increasingly greater role or even a dominant role in determining the Eh of the soil when the Eh falls to zero or negative value. An auxiliary experiment (Fig. 1-10) showed this clearly. In the treatment, with the addition of green manure, the Eh of the soil fell to -263 mV one day after submergence. Since at this time the oxygen had already disappeared and the amount of ferrous iron was still very small, the Eh of the soil should be determined only by organic oxidation-reduction systems. On the other hand, in the check treatment, the Eh of the soil can maintain a high level within two days of submergence due to the low content of organic matter of the original red soil.

It is interesting to consider the significance of the iron system in determining the Eh of paddy soils. An iron system is a reversible system at the surface of a platinum electrode. In many cases the concentration of ferrous iron may be quite high. Thus iron may function as a determining system for the electrode potential. However, the transformation of iron in soils is primarily controlled by oxygen and organic oxidation-reduction systems. If the rate of change in oxidation-reduction condition is very fast, the transformation of the iron system may lag behind the change in the Eh of the soil. For instance, in the

1.2 Eh in relation to chemical equilibria in paddy soils

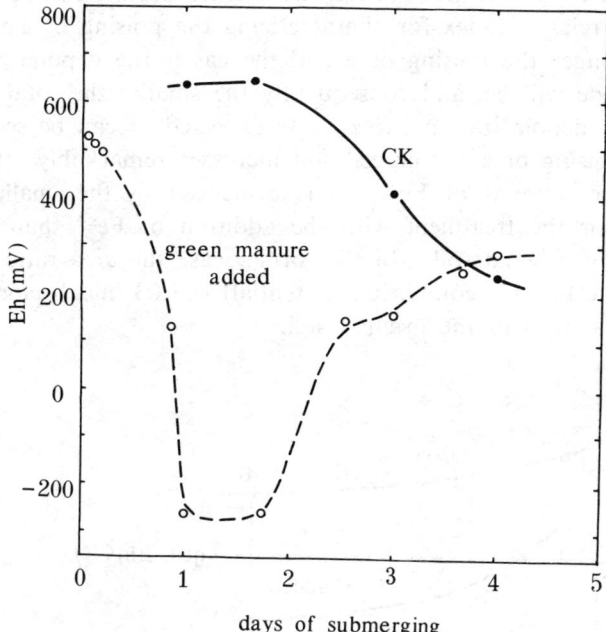

Fig. 1-10 Influence of green manure on change of soil Eh (red soil) (courtesy of S. Z. Pan)

cultivated layer of the "White soil" with an Eh of 96 mV on the 6th day of submergence, the content of ferrous iron was only 0.7 m.e. per 100g of soil, whereas after drying the submerged soil the content of ferrous iron maintained a high level of 10.7 m.e. per 100g, although the Eh had risen to 590 mV. Therefore, it may be concluded that although iron may be the determining system with respect to electrode potential it is not the determining system with respect to the Eh of the soil *per se*. It may be regarded as a kind of buffering system.

From what has been said above it is clear that the Eh of paddy soils is determined not always by the same system, but by different systems under different conditions.

1.2.4 Poising

The poising of a soil refers to the ability of the soil to resist the change in the Eh upon the addition of a small amount of oxidant or reductant. According to Equation (1-11), the concentration of oxidant or reductant is closely related to the poising of the system, although it has no direct proportional bearing on the Eh. At a constant concentration of the oxidation–reduction system, the poising will be strongest when the ratio of oxidant to reductant is close to 1. When a chemical reaction occurs among several oxidation–reduction systems, the change in Eh will be controlled by the system which is highest in concentration due to its strongest poising.

It is possible to use the area between the anodic and the cathodic depolarization curve as a relative index for characterizing the poising of a soil. This is because the stronger the poising of a soil the easier the depolarization of the platinum electrode will be, and consequently the smaller the total area will be between the two depolarization curves. As expected, it can be seen from Fig. 1-11 that the poising of a submerged soil increases remarkably after the addition of a certain amount of Fe^{2+}, as is evidenced by the smaller total area within ABCD for the treatment with the addition of Fe^{2+} than that within A'B'C'D' for the original soil In the former case the area ratio of ABFE to DCFE (EF is the line for equilibrium potential) is 1.03, much closer to 1 than 2.59 for the area ratio of the original soil.

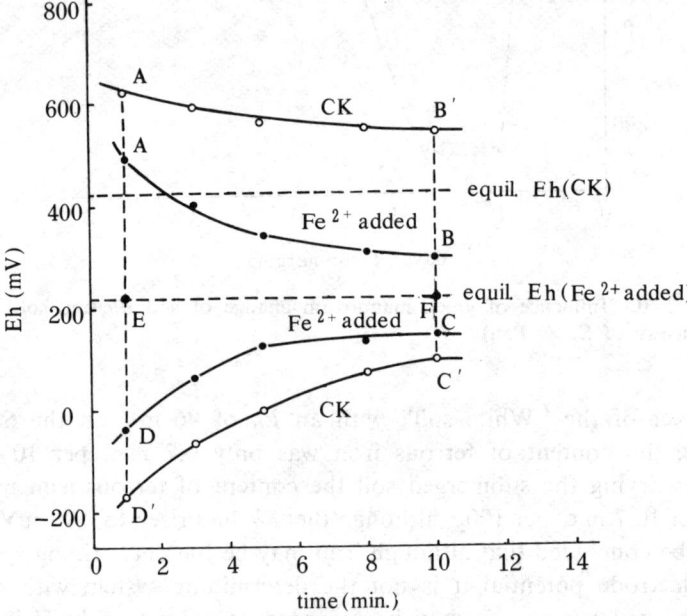

Fig. 1-11 Effect of Fe^{2+} on depolarization curve (submerged red soil)

In Table 1-6 are shown the results for some paddy soils. Soils with stronger poising are characterized by a smaller total area between two depolarization curves and an area ratio closer to 1 for the two curves. Besides, it was found that the stronger the poising of a soil, the more consistent the Eh between those measured by the extrapolation method and the equilibrium value.

The concept of poising of the soil is very useful. However, up to the present research work in this respect is still very rare.

1.2.5 Gradation of oxidation-reduction status according to Eh

Although there are some theoretical and technical problems in the measurement of the Eh of the soil, it is possible to roughly grade the oxidation-reduction status of soils according to the Eh. This is because the Eh is the most impor-

1.2 Eh in relation to chemical equilibria in paddy soils

TABLE 1–6 Relationship between poising and character of depolarization curve

Soil no.	Total area*	Area ratio	Eh (mV)			Poising
			Extrapolated (A)	Measured (B)	(A)–(B)	
14	12.9	1.01	3	4	–1	Strong
22	13.5	1.14	62	62	0	Strong
10	14.3	1.17	31	32	–1	Strong
1	15.7	1.25	85	82	3	Strong
17	16.8	1.55	11	15	–4	Moderate
32	16.6	3.37	–79	–85	6	Weak
4	19.6	2.70	–35	–42	7	Weak
18	24.9	3.37	–11	–20	9	Weak
12	27.2	1.32	372	380	–8	Weak
8	28.4	4.26	–112	–123	11	Weak

* Arbitrary relative unit

tant index for characterizing the oxidation–reduction strength of the soil, and at the same time if compared with a wide range of variation in the Eh among soils, errors in measurements are usually not so serious as to cause the measured Eh to lose its relative significance. Based on its relationships with other soil properties and plant growth the oxidation–reduction status of soils may be classified into four categories, namely oxidizing, weakly reducing, moderately reducing and strongly reducing. The corresponding Eh values are higher than 400, 400–200, 200–(–100) and lower than –100 mV respectively (Table 1–7).

TABLE 1–7 Gradation of redox status of soils

Redox status	Eh range (mV)	Reaction	Plant growth
Oxidizing	>400	O_2 predominant, materials in oxidized form	Beneficial to upland crops, not suitable for rice
Weakly reducing	400–200	O_2, NO_3^- and Mn^{4+} reduced	Normal growth of rice, upland crops affected
Moderately reducing	200–(–100)	Fe^{3+} reduced, organic reducing substances present	Upland crops affected
Strongly reducing	<(–100)	CO_2 and H^+ reduced	Rice affected by reducing substances

1.3 HETEROGENEITY IN OXIDATION–REDUCTION POTENTIAL IN PADDY SOILS

Owing to the nonuniform distribution of oxygen, organic matter and some inorganic solid materials in the soil, the micro–regional difference in oxidation–reduction regimes of paddy soils is a commonly encountered phenomenon. The diffusion rate of oxygen in the interstitial water during submergence is less than one thousandth of that in the air, and as a consequence the soil mass as a whole may be considered as oxygen–deficient. However, in the uppermost part of the cultivated layer in contact with the surface water containing much dissolved oxygen, the content of oxygen may be higher than that in the lower part, thus inducing a micro–regional difference in the Eh near the interface. At different parts of aggregates and interstices there may be a difference in Eh due to differences in the contacting condition with the air. The Eh at the surface of soil particles may be different from that of the liquid phase. Besides, micro–regional differences in Eh between the root–zone of plants and the bulk soil are also distinct. These heterogeneities in the Eh will be discussed in the following sections.

1.3.1 Oxidizing layer and reducing layer near water–soil interface

In the upper part of submerged soils the concentration of oxygen is generally higher than that of the lower layers due to the diffusion of oxygen from the surface water. This induces the formation of an oxidizing layer which is lighter in color than the underlying layers. The thickness of this oxidizing layer is related to the criterion on which the distinction is based. It was found that the Eh was a better index in this respect. It can be seen from a measurement on a sandy paddy soil shown in Fig. 1–12 that the thickness of this layer is about 8 mm if the Eh is used as the index. Usually 200—250 mV at pH 7 may be considered as the demarcation point for distinguishing the oxidizing layer and the reducing layer. If S^{2-} concentration is used as the index of demarcation, the thickness is about 10 mm (Fig. 1–12).

In the oxidizing layer, in addition to the high content of oxygen and high Eh, some elements are present in their oxidized form such as NO_3^-, Fe^{3+}, Mn^{4+} and SO_4^{2-}. On the contrary, in the reducing layer the Eh is low as a result of a high content of reducing substances (Table 1–8).

The rate of formation and the thickness of the oxidizing layer are significant in agricultural practice, because the properties of this layer are related to the loss of nitrogen and the transformation of some other nutrients in the soil. Morphologically, the thickness of this layer ranges from 0.1 mm to more than 10 mm, depending on soil conditions. The thickness of this layer is in essence a reflection of the dynamic equilibrium between the oxidant (O_2) and the reductant of the soil. When the amount of easily decomposable organic matter is high the thickness of the oxidizing layer will be small (Table 1–8). This layer may even be absent if the amount of dissolved oxygen in water is not sufficient to oxidize

1.3 Heterogeneity in oxidation-reduction potential in paddy soils

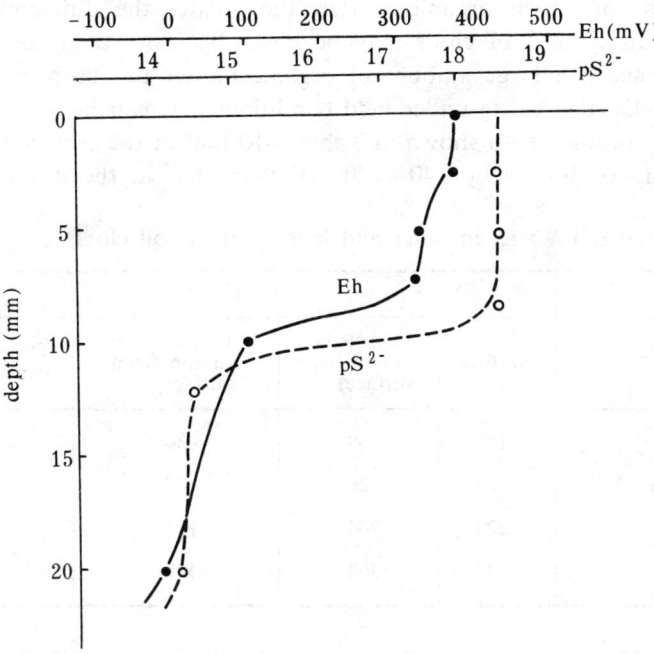

Fig. 1-12 Eh and pS^{2-} in oxidizing layer and reducing layer (sandy paddy soil)

TABLE 1-8 Difference in properties between oxidizing layer and reducing layer (sandy paddy soil)

Treatment	Layer	Eh (mV)	Water-soluble ions			S^{2-} (M)	O_2 (mg/l)
			Fe^{2+}	Mn^{2+}	NO_3^-		
CK	Oxidizing (6 mm thick)	320	1	1.5	1.2	Undetectable	3.21
	Reducing	68	14	5.1	Undetectable	5×10^{-17}	tr.
Dried milk vetch	Oxidizing (3 mm thick)	225	1.5	3.7	0.51	Undetectable	1.55
	Reducing	−79	57	9.8	Undetectable	3.5×10^{-15}	tr.

all of the reducing substances formed in the soil. On the other hand, if the amount of easily decomposable organic matter is very low and the soil is very permeable, as for instance in the case of very infertile sandy soils, there will be no distinguishable differentiation between the oxidizing layer and reducing layer due to the weak development of reducing condition during submergence.

1.3.2 Inner and outer parts of soil clods and cleavage surface

The difference in Eh between the inner and the outer part of soil clods is caused by the difference in aeration. It is seen from Table 1-9 that the larger the amount of added organic matter, the greater the difference in Eh. The Eh in the inner part of clods may be lower by more than 200 mV than that at the surface if a large amount of organic matter has been added. This difference in Eh also exists under field conditions. It can be seen from measurements on a drained field shown in Table 1-10 that at the surface of the soil clod the Eh may be higher by 140—230 mV than that in the interior of the clod.

TABLE 1-9 Eh in outer and inner part of soil clod[8]

Content of milk vetch	Eh (mV)			
	Surface	Intermediate (3 mm from surface)	Center (5 mm from surface)	Difference between extremities
High	188	68	-34	222
Medium	207	124	67	140
Low	228	204	193	35
Check	437	369	328	109

TABLE 1-10 Eh in different parts of soil clods with different water content (paddy soil from yellow-brown soil)

Water content (%)	Eh (mV)*			
	Surface	3 mm from surface	Center	Difference between extremities
23.5	157	61	-70	227
30.1	52	-36	-90	142

* Mean value of 20—30 determinations

At the cleavage surface the soil is less reduced during the submerging period due to the influence of percolating water containing a little dissolved oxygen. After drainage the Eh rises rapidly to higher than 400 mV, although in the interior of the soil solum the Eh may maintain a low level for some time.

1.3.3 Centrifugate, suspension and bulk soil

It has been frequently observed that if Eh measurements are made for the bulk soil and its equilibrium solution separately the results will be different. It can be seen from Table 1-11 that for soils with an Eh higher than 500 mV the measured Eh of the bulk soil is higher than that of the centrifugate, whereas for reduced soils it is lower than that of the centrifugate. An acid sulfate soil (pH 2.15) is an exception in which the two Eh values are the same.

It is not known why the measured Eh is different when the platinum electrode is in contact with the bulk soil and with the solution.

TABLE 1-11 Difference in *Eh* among centrifugate, suspension and bulk soil of some paddy soils and their parent soils

Paddy soil or parent soil	Locality	*Eh* (mV)			
		Bulk soil (A)	Suspension*	Centrifugate (B)	(A) – (B)
Acid sulfate	Fujian	586	555	542	+44
With rusty water	Fujian	585	520	571	+14
Red soil	Jiangxi	568	565	560	+3
Red sandy soil	Jiangxi	552	540	525	+27
Paddy from red sandy soil	Jiangxi	455	—	475	−20
Paddy from yellow-brown soil	Jiangsu	391	415	421	−30
Acid sulfate	Guangdong	364	364	364	0
Paddy from red soil	Jiangxi	78	110	190	−112
Gleyed	Fujian	73	—	275	−202
Loamy	Jiangxi	51	—	188	−137
Clayey	Jiangsu	50	—	220	−170
Paddy from red soil	Zhejiang	−277	—	−250	−27

* Determined 1—2 days before

1.3.4 Root–zone and bulk soil

Rice roots can secrete oxygen to the surrounding soil, thus causing the *Eh* of the root–zone to be higher than that of the bulk soil. On the contrary, when upland crops are grown the organic reducing substances secreted by roots enable the *Eh* f the root–zone to be lower, as already demonstrated by many experiments and field measurements[1,6]. This question will be dealt with in more detail in Chapter 10.

1.4 OXIDATION–REDUCTION POTENTIAL OF PADDY SOILS

1.4.1 Oxidation–reduction potential of paddy soils with different water regimes

The oxidation–reduction potential of paddy soils is profoundly affected by the water regime of the soil. Based on the depth of water–table, the water regime may be classified into three main types. In Fig. 1-13 are shown some of the examples in Tai Lake Plain.

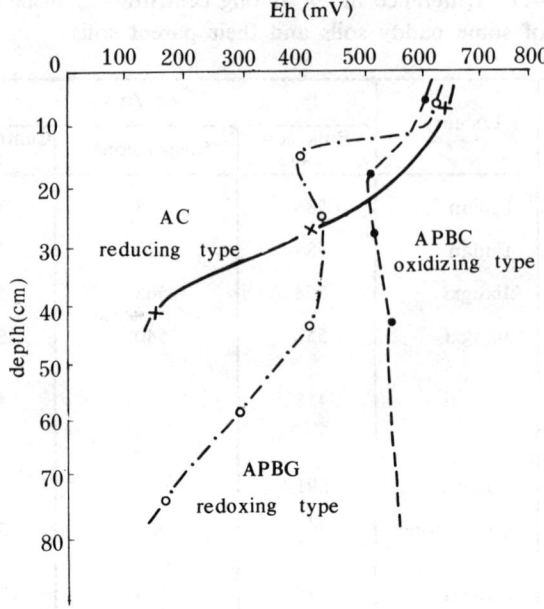

Fig. 1-13　*Eh* of paddy soil profiles with different water regimes (Jiangsu)[3]

A-C and A-P-B-C type: In the A-C subtype of paddy soil, the duration of submergence is short and the water-table is low. The whole profile is mainly governed by oxidation processes, except the cultivated layer. In the A-P-B-C subtype of paddy soil, the soil solum within one meter is also not affected by ground water. This type of paddy soil may be called oxidizing paddy soil.

A-P-B-G type: In the A-P-B-G type of paddy soil, there is a glei horizon in the lower part of the one meter solum due to the presence of ground water. The *Eh* of the glei horizon is usually lower than 250 mV. Within a certain distance from the glei horizon the *Eh* is also lower than the upper horizons due to a higher water content. This type of paddy soil may be called redoxing paddy soil.

A-G and A-P-G type: The water-table in these soils is very high, and most of the whole profile is under a strongly reducing condition. There are also gleyed paddy soils with ground water-table approaching the soil surface throughout the year (Fig. 1-14). The *Eh* of the soil is 100—200 mV or even a negative value. This type of soil may be called reducing paddy soil.

1.4.2　Dynamics of oxidation-reduction potential in the profile

The seasonal dynamic regime of oxidation-reduction potential of paddy soils may be illustrated by taking the cultivated layer of the A-P-B-C type of soil as an example. The *Eh* generally maintains a value of 450—650 mV before submerging. After submerging it declines rapidly to -200 to $+100$ mV at the period

1.4 Oxidation-reduction potential of paddy soils

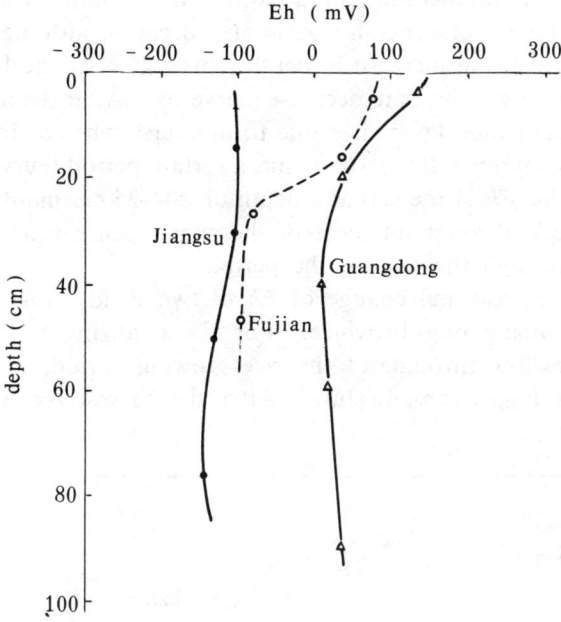

Fig. 1-14 Eh of gleyed paddy soil profiles[3]

of intensive decomposition of organic matter (about the 3rd—10th day). In extreme cases where a large amount of easily decomposable green manure has been added the Eh may be as low as -300 mV. After this period the Eh rises to a steady value of 0—200 mV. Before the harvest of rice, the Eh returns to higher than 450 mV again due to the drainage of the field.

In Fig. 1-15 is shown the change of Eh in the profile of an oxidizing paddy soil ("White soil"). Before submerging, the Eh of the whole profile is higher

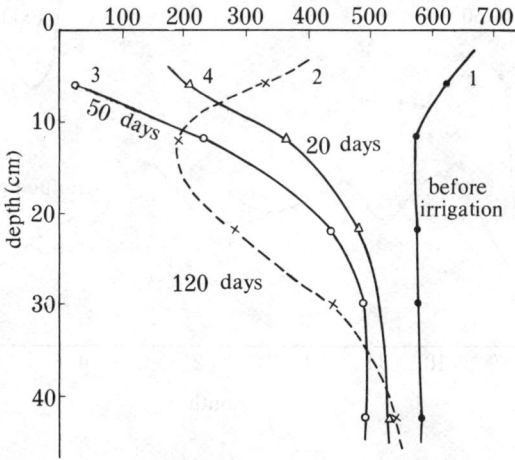

Fig. 1-15 Dynamics of Eh in an oxidizing paddy soil profile (Jiangsu, "White soil")[3]

than 500 mV (curve 1). After submerging, in addition to a sharp drop of Eh in the cultivated layer, the Eh of other horizons also declines, although to a progressively lesser extent with the increase in depth (curve 2). At the latter stage the Eh of various horizons continues to decrease (curve 3). After the milky stage of rice the Eh of the cultivated layer rises due to drainage, whereas in the underlying horizons it still maintains a low level within a certain period (curve 4). It is interesting to note that the Eh of the soil at a depth of 20—25 cm maintains a level of higher than 400 mV throughout most of the submergence time, and the Eh below 40 cm is quite high throughout the period.

In Fig. 1-16 is shown the seasonal change of Eh of two paddy soils with different water regimes in Guangdong Province. For the oxidizing soil, the Eh of the cultivated layer is low throughout the rice-growing period, and in the underlying horizons it is high except in July. After the harvest the whole

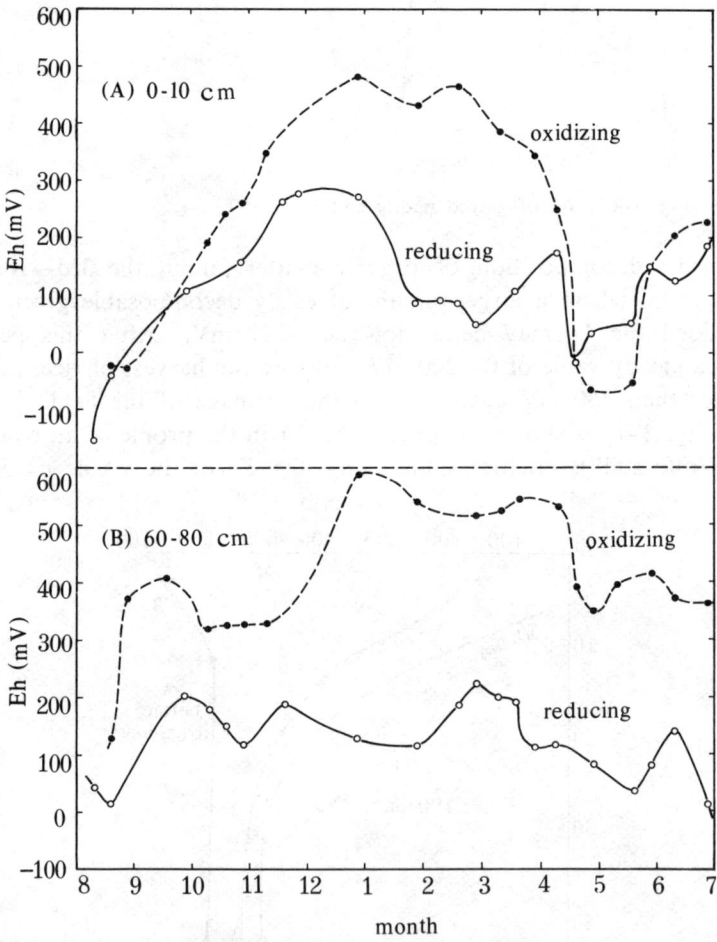

Fig. 1-16 Seasonal change of Eh of two paddy fields with different water regimes (Guangdong) (courtesy of Z. C. Chen and Z. T. Gong)

1.4 Oxidation-reduction potential of paddy soils

profile is under an oxidizing condition. For the reducing paddy soil the Eh is low throughout the year, especially in the summer season.

1.4.3 Oxidation–reduction potential of representative paddy soils

From the viewpoint of zonal variation in oxidation–reduction potential of the soil it is very difficult to sum up a general regularity, for the oxidation–reduction potential of a soil is chiefly affected by water regime and organic matter status of the soil, and usually there is no clear zonal pattern for the latter two factors. However, inasmuch as the Eh of a redox system is affected by pH of the medium, it can frequently be observed that if comparisons are made for various types of paddy soil there is a correlation between the Eh and the pH in the non-irrigated season. In Fig. 1-17 are shown some of the examples in this regard. The Eh of paddy soils derived from red soils with a pH of 4.6—5.1 is 680—720 mV. The corresponding figures for "White soil" in Jiangsu Province are 6.2—6.5 and 550—610 mV respectively, and those for paddy soils derived from yellow-brown soil are 6.5—7.0 and 515—530 mV respectively. For paddy soils derived from drab soil in the Beijing region with a high pH (7.2—8.0) due to the presence of calcium carbonate the Eh is usually 450—500 mV. If we consider that for these non-submerged soils the oxidation–reduction of the substratum is primarily governed by oxygen it is interesting to note that the magnitude of change in the Eh with the change in pH is about 60 mV per 1 pH unit, which is consistent with theoretical expectations.

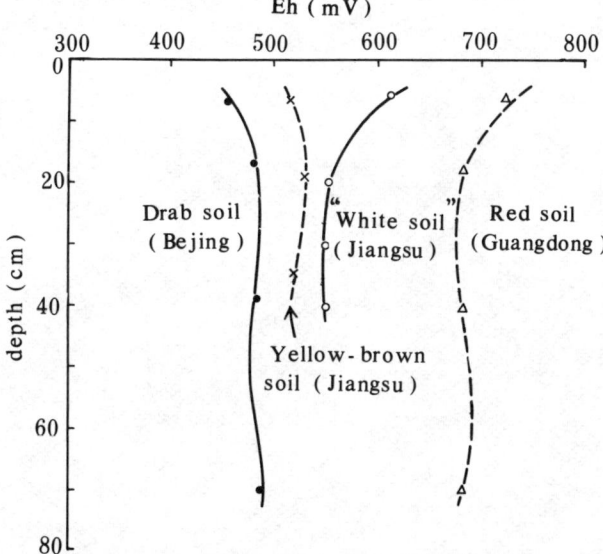

Fig. 1-17 Eh of representative paddy soil profiles[1,3]

REFERENCES

(1) Institute of Soil Science, 1961. Soil Environment of High-yield Rice. Chapter 6. Science Press, Beijing.

(2) Yu Tian-ren et al., 1976. Electrochemical Properties of Soils and Their Research Methods (revised ed.). Chapter 11. Science Press, Beijing.
(3) Institute of Soil Science, 1978. Soils of China. Part B, Chapter 10. Science Press, Beijing.
(4) Yu Tian-ren, Zhang Xiao-nian et al., 1980. Electrochemical Methods and Their Applications in Soil Research. Chapter 11. Science Press, Beijing.
(5) Yu Tian-ren and Li Sung-hua, 1957. Studies on oxidation-reduction processes in paddy soils. I. Factors affecting oxidation-reduction potential. Acta Pedologica, **5**: 97–110.
(6) Yu Tian-ren and Li Sung-hua, 1957. Studies on oxidation-reduction processes in paddy soils. II. Mutual influences between soil and plants. Acta Pedologica, **5**: 166–174.
(7) Yu Tian-ren, Xie Jian-chang and Yang Guo-zhi, 1959. Eh-determining systems in paddy soils. Science Bulletin, **6**: 205–206.
(8) Liu Zhi-guang and Yu Tian-ren, 1963. Studies on electrochemical properties of soils. II. Applications of micro-electrodes in soil research. Acta Pedologica, **11**: 160–169.
(9) Liu Zhi-guang, 1981. Determination of redox potential (Eh) by depolarization curve method. in "Proc. Symp. Paddy Soil", pp. 258–261. Science Press, Beijing. (in English)
(10) Ponnamperuma, F. N., 1972. The chemistry of submerged soils. Adv. Agron., **24**: 29–96.

CHAPTER 2

REDUCING SUBSTANCES

DING CHANG-PU AND LIU ZHI-GUANG

The decline of oxidation–reduction potential in paddy soils after submerging is due to the production of reducing substances. Following the strengthening of reducing conditions, the oxidized forms of various redox systems present in the soil are transformed to their reduced forms consecutively, or the ratio of the oxidized forms to the reduced forms diminish, as a result of which the oxidation–reduction potential is lowered. The oxidation–reduction potential discussed in the last chapter expresses the degree of oxidation or reduction. Therefore, it is an intensity factor. It also reflects, to some extent, the relative proportion of oxidizing or reducing substances, which is a capacity factor. In this chapter this capacity factor of oxidation–reduction properties of the soil will be discussed.

2.1 CHARACTERIZATION OF REDUCING SUBSTANCES OF THE SOIL

Reducing substances of the soil may be characterized according to their reducing intensity or chemical activity, which are discussed as follows.

2.1.1 Reducing intensity

The reducing intensity of a reducing substance represents its ability to donate or accept electrons. It is possible to characterize the reducing intensity and the quantity of reducing substances which possess a given reducing intensity by making the soil be in contact with a metallic conductor (electrode) capable of transferring electrons, and then getting the current–voltage curve[2, 4]. The ability of the electrode to accept electrons is controlled by an appropriate potential applied externally. According to the principle of voltammetry the reducing substance can produce an anodic current at a carbon electrode due to its oxidation under an applied potential. Under a constant experimental condition and with a constant electrode area, the current (diffusion current, i_d) should be proportional to the concentration (C) of the reducing substance according to the following equation:

$$i_d = KC \tag{2-1}$$

Theoretically, different reducing substances should have their characteristic half-wave potentials, and thus it is possible to distinguish them precisely through the control of applied potential. In practice, however, it is difficult to do so, because the differences in half-wave potential among some reducing substances are not large enough, the electrochemical reaction of some reducing substances at the electrode is not reversible, and there may also be interactions occurring among them or with other substances. Nevertheless, for practical purposes this method of characterization is usable, as is shown in the following section.

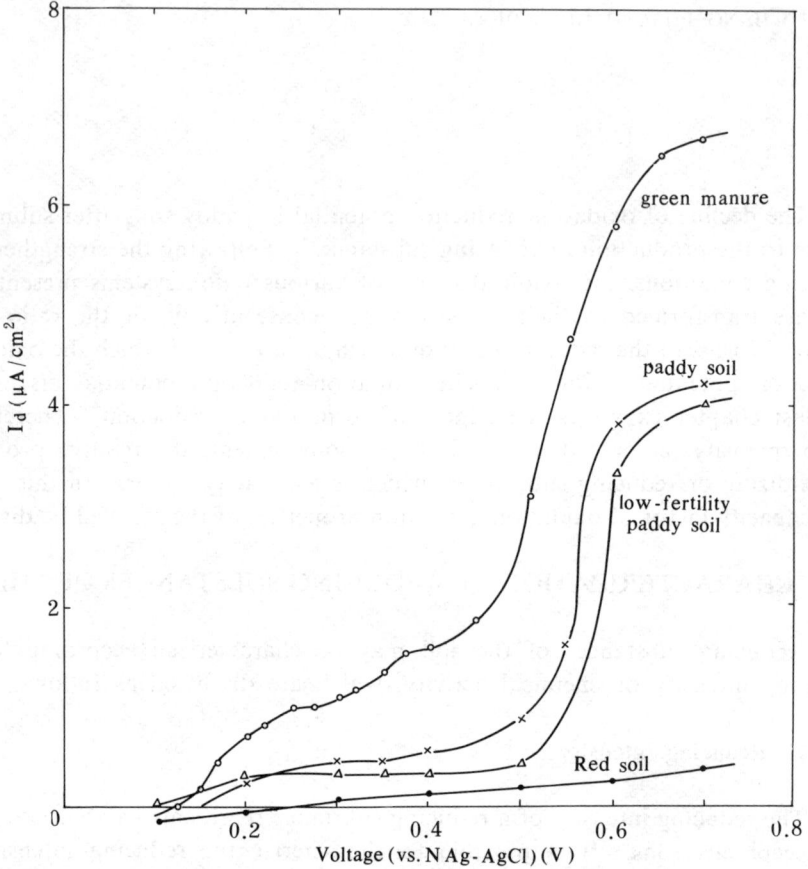

Fig. 2-1 I-V curves for reducing substances on graphite electrode[8]

In Fig. 2-1 are shown the current–voltage (I–V) curves for three soil suspensions and an incubation solution of green manure. It can be seen that the diffusion current increases with the increase of applied voltage. These curves are representative of a large number of reduced soils. Since there are two plateau regions at about 0.3—0.4 V and 0.6—0.7 V respectively, it is convenient to use the diffusion current at 0.35 V as an index for expressing the quantity of strongly reducing substances, and that at 0.7 V for expressing the total amount of strongly and weakly reducing substances; their difference will represent the quantity of weakly reducing substances.

It can also be seen from the figure that for a red soil with an organic matter content of less than 0.5% the amount of reducing substances is very low even under submerged conditions, whereas for the decomposition products of green manure there are large amounts of reducing substances. Between the two paddy

2.1 Characterization of reducing substances of the soil

soils the amount is lower in the low–fertility soil.

The diffusion current is caused by all the substances capable of participating in electrode reactions. It can be seen from the experimental results shown in Fig. 2–2 that at pH 6—7 (the usual pH range in reduced paddy soils) Fe^{2+} and Mn^{2+} are included in the strongly and the weakly reducing substances respectively. Since the contents of S^{2-} and H_2S in soil solution are usually low, the other part of diffusion current should be chiefly caused by organic reducing substances.

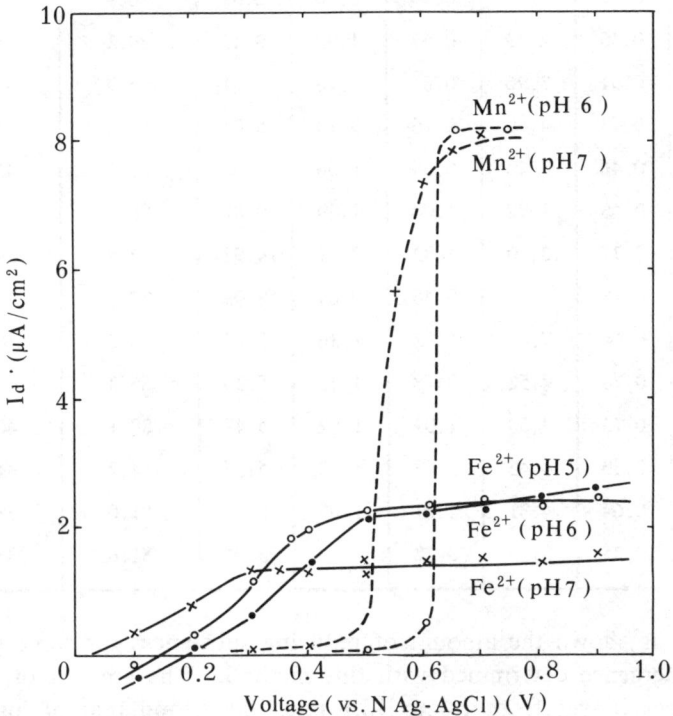

Fig. 2–2 I–V curves for Fe^{2+} and Mn^{2+} [8]

2.1.2 Chemical activity

It is also possible to characterize reducing substances according to their activities with respect to chemical reactions toward oxidation reagents. The reducing substances may be extracted with a 0.1 M $Al_2(SO_4)_3$ solution of pH 2.5, and then titrated with $KMnO_4$ at room temperature to determine the active reducing substances, and oxidized with $K_2Cr_2O_7$ at 90—95°C to determine the total reducing substances. The difference between the two is taken as the amount of inactive reducing substances. This method of characterization according to the liability of oxidation is quite convenient for practical purposes, although the absolute figure of results is conditional.

TABLE 2-1 Active and inactive reducing substances in paddy soils[7]

Soil name	Locality	Reducing substances (m.e. / 100g)					Ratio	
		Active		Inactive (C)	Total organic (A+C)	Sum (B+C)	$\frac{(B)}{(B+C)} \times 100$	$\frac{(A)}{(A+C)} \times 100$
		Organic (A)	Total (B)					
Beishan	Nanjing	0.36	3.30	0.56	0.92	3.86	85.5	39.1
Huangni	Wuxi	0.06	1.69	0.40	0.46	2.09	80.9	13.0
Huangni	Changsha	0.35	2.73	0.67	1.02	3.40	80.3	34.3
Huangni	Bolo	0.51	3.96	0.65	1.16	4.61	85.9	44.0
Huangni	Bolo	0.98	4.69	1.16	3.14	5.85	80.2	45.8
Wusha	Zixi	0.46	3.40	0.60	1.06	4.00	85.0	43.4
Heisha	Anyi	0.45	9.78	1.04	1.49	10.82	90.4	30.2
Heisha	Anyi	0.77	13.10	1.82	2.59	14.92	87.8	29.7
Guoba	Luocheng	0.18	1.57	0.46	0.64	2.03	77.3	28.1
Qingzi	Ningpo	0.14	7.62	1.22	1.36	8.84	86.2	10.3
Yashi	Xinghua	0.70	4.52	0.78	1.48	5.30	85.3	47.3
Hungtu	Xinghua	0.73	4.38	1.09	1.82	5.47	80.1	40.1
Chitu	Xuiwen	0.59	4.43	0.83	1.42	5.26	84.2	41.5
Heinian	Nanning	0.64	4.31	1.01	1.65	5.32	81.0	38.8
Heinian	Nanning	0.45	3.19	0.72	1.17	3.91	81.6	38.5

In Table 2-1 is shown the amount of reducing substances for some paddy soils during submergence determined with this method. The amount of active reducing substances ranges from 1.6 to 13.1 m.e./100 g, and that of inactive reducing substances 0.4—2.2 m.e./100 g. The former accounts for more than 60% of the total reducing substances. The inactive portion should be mainly organic matter which is difficult to oxidize. It can be seen from the table that the active organic reducing substances generally account for more than 80% of the total organic reducing substances. If consideration is given with reference to Fig. 2-1, it may be assumed that most of these active organic reducing substances are capable of participating in electrode reactions.

In the characterization mentioned above the active reducing substances comprise both Fe^{2+} and organic matter. Since Mn^{2+} does not participate in the chemical reactions mentioned above, it is not included in the reducing substances determined by the proposed method. If necessary, it may be determined separately.

2.1.3 Kinds and properties

Reducing substances of the soil consist of both inorganic and organic species.

2.1 Characterization of reducing substances of the soil

Inorganic reducing substances are chiefly composed of Fe^{2+}, Mn^{2+} and S^{2-}, which will be discussed in later chapters. In this section is given a brief account of some properties of organic reducing substances.

Active organic reducing substances are a group of substances which are readily extractable with water and can rapidly participate in chemical oxidation–reduction reactions. They can also participate in electrode reactions. However, the mechanisms of these electrode reactions are frequently irreversible or semi–reversible.

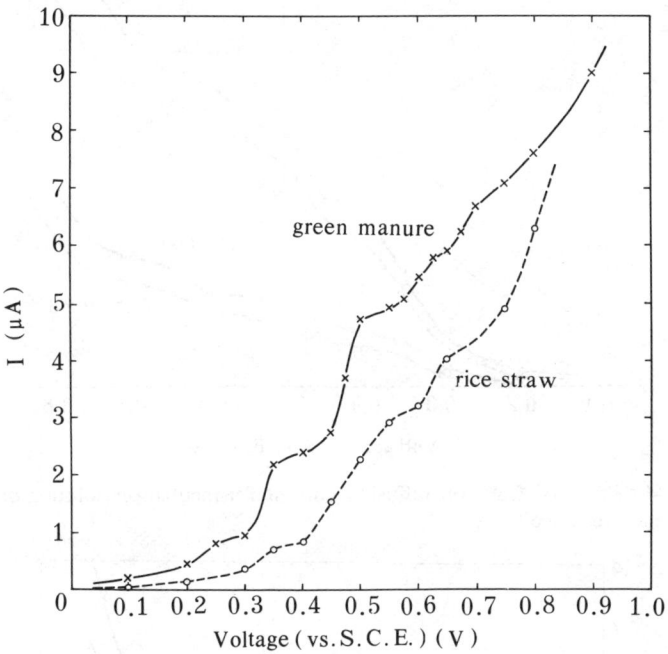

Fig. 2–3 I–V curves for decomposition products of green manure (carbon-paste electrode)

In Fig. 2–3 are shown the I–V curves for the incubation solutions of two green manures. According to their electrochemical behavior, the decomposition products of milk vetch can be distinguished as several groups with "half–wave" potentials of 0.22, 0.33, 0.47 and 0.65 V (vs. saturated calomel electrode) respectively. The decomposition products of rice straw are characterized by smaller amounts and indistinguishable wave–forms.

The so–called "half–wave" potential for each group has no definite qualitative meaning as in classical voltammetry.

It was shown on oscillograms that for milk vetch there were three groups of reducing substances with half–wave potentials of 0.04, 0.23 and 0.70 V at the middle stage of its decomposition.

Some of the organic reducing substances probably possess the ability of complexation with some metal ions[3]. It can be seen from Fig. 2–4 that for the incubation solutions of green manure the addition of Cu^{2+} leads to the decrease of the diffusion current, particularly for those components with high half-

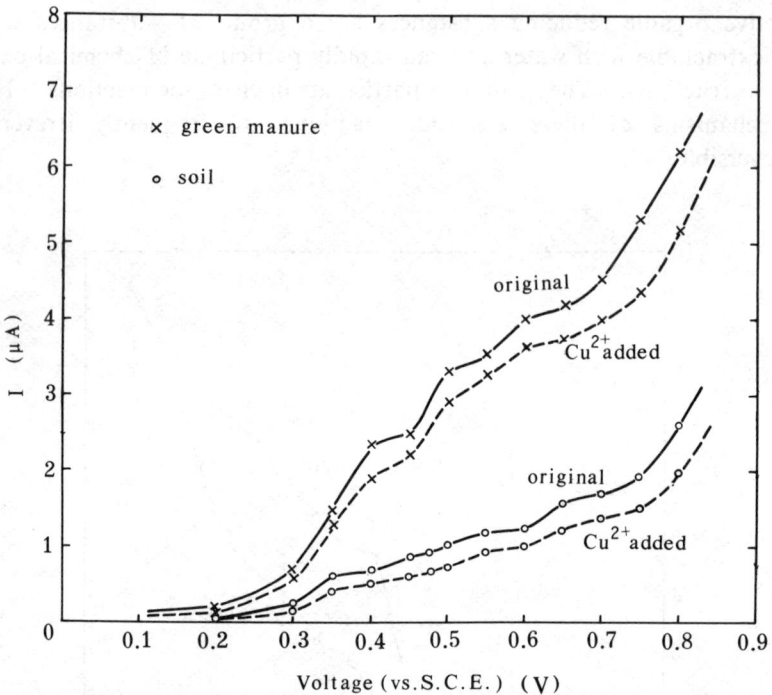

Fig. 2-4 Effect of Cu^{2+} on diffusion current for incubation solution of green manure of soil

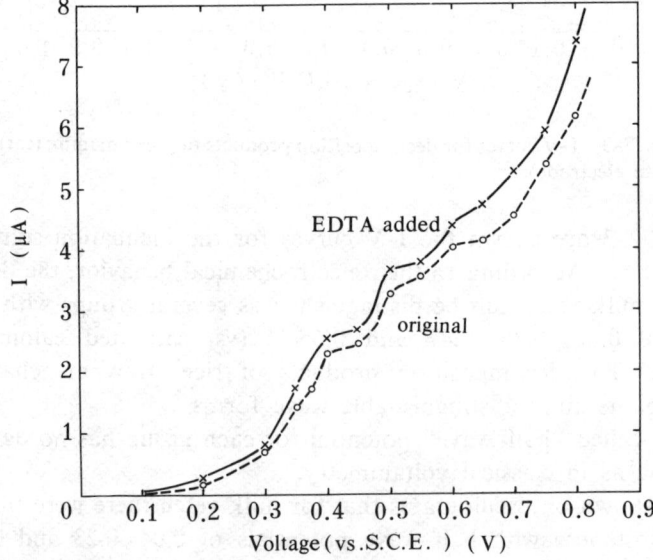

Fig. 2-5 Effect of EDTA on diffusion current for incubation solution of green manure

wave potentials. On the other hand, the addition of EDTA induces an increase in diffusion current (Fig. 2–5), due probably to the liberation of some organic substances originally complexed by metal ions of the solution.

2.2 FACTORS AFFECTING THE AMOUNT OF REDUCING SUBSTANCES IN PADDY SOILS

2.2.1 Organic matter

Organic matter affects the amount of reducing substances either directly or indirectly. Organic matter is the principal source of electrons of the soil, and can produce reducing substances during its decomposition. These organic reducing substances can further undergo electron exchange with ferric and manganic oxides and sulfates, thus reducing them into Fe^{2+}, Mn^{2+} and S^{2-}. An example of such effect is shown in Fig. 2–6. The diffusion current for the original red soil without the addition of green manure is only 0.39 $\mu A / cm^2$ at +0.7 V, close to residual current. The current in the green manure treatment increases to 2.94 $\mu A / cm^2$, due apparently to the increase in the amount of reducing substances caused by the decomposition of organic matter. Therefore, it was observed that for paddy soils with higher organic matter content

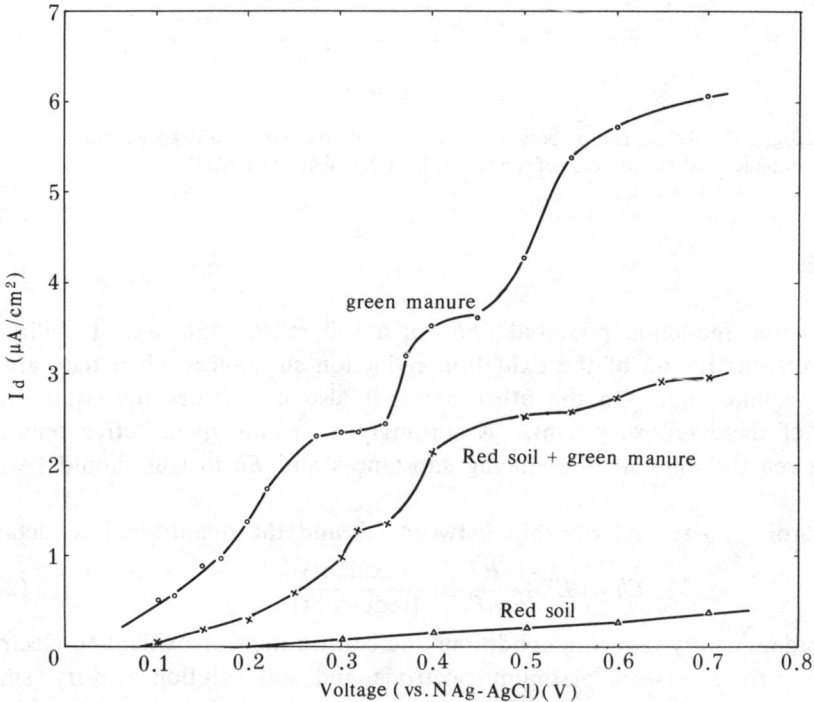

Fig. 2–6 Influence of organic matter on amount of reducing substances in the soil (electrochemical method

the amount of reducing substances during submergence determined by chemical method was usually higher. In Fig. 2–7 is shown the relationship between the total amount of reducing substances during submergence and the organic matter content of paddy soils. It must also be admitted that the points are not on a straight line in the figure. This is due to differences in the composition of organic matter and the liability to decomposition in different soils.

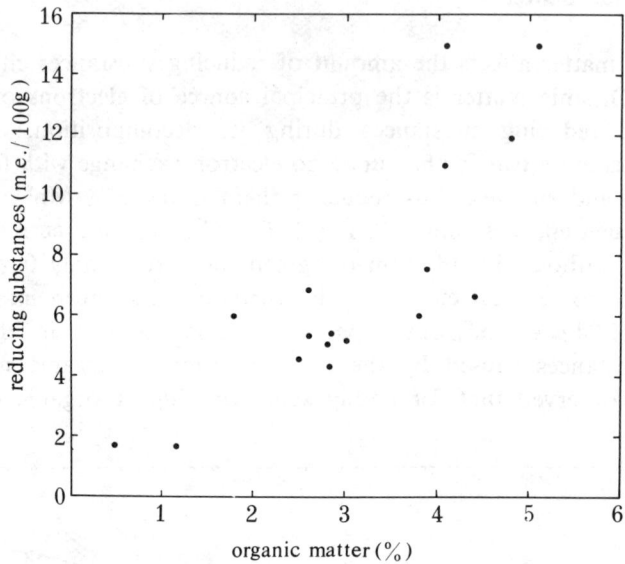

Fig. 2–7 Relationship between amount of reducing substances and organic matter content of paddy soils (chemical method)[1]

2.2.2 Eh

Oxidation–reduction potential *(Eh)* of a soil reflects the overall ability to accept electrons for all of the oxidation–reduction substances when they are in chemical equilibrium. On the other hand, it also determines the equilibrium position of these redox systems. Accordingly a definite quantitative relationship between the amount of reducing substances and *Eh* in soil should be anticipated.

According to the relationship between *Eh* and the quantity of reductant:

$$Eh = E^0 + \frac{RT}{nF} \ln \frac{(\text{oxidant})}{(\text{reductant})} \qquad (2\text{–}2)$$

under predominantly reducing conditions the contribution of oxidant to electron exchange current between platinum electrode and soil solution is very small and can be incorporated into the E^0 term due to the low concentration of oxidant, and we can get:

$$Eh = E^{0\prime} - a \log (\text{reductant}) \qquad (2\text{–}3)$$

It can be seen from the equation that there should be a linear relationship

2.2 Factors affecting the amount in paddy soils

between the *Eh* and the logarithm of the quantity of reductant. In such a special case the amount of reducing substances can play a predominant role in the mixed potential as determined with a platinum electrode. This can be illustrated by the following examples.

Qualitatively, a large number of determinations showed that the higher the amount of reducing substances in a paddy soil, the lower the oxidation–reduction potential. It can be seen from the potentiometric titration curves of Fig. 2–8 that for two soils with an *Eh* of 30 and 140 mV the amounts of reducing substances were 7.4 and 4.6 m.e. / 100 g respectively.

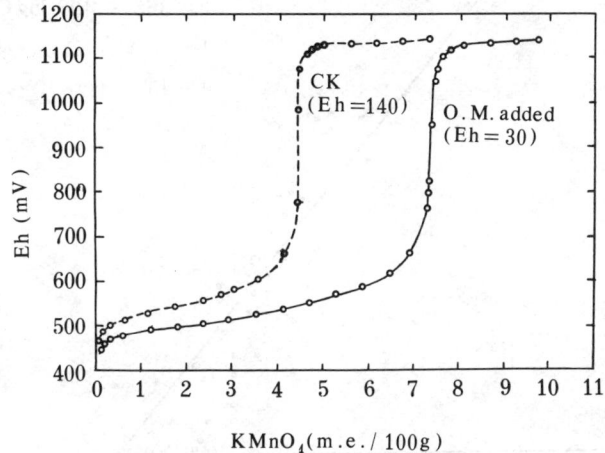

Fig. 2–8 Potentiometric titration curves of reducing substances in paddy soil[6]

Quantitatively, it has been found that between the logarithm of the amount of reducing substances determined by chemical method for some submerged paddy soils and Eh_7 (*Eh* corrected to pH 7) there is a fairly good negative cor-

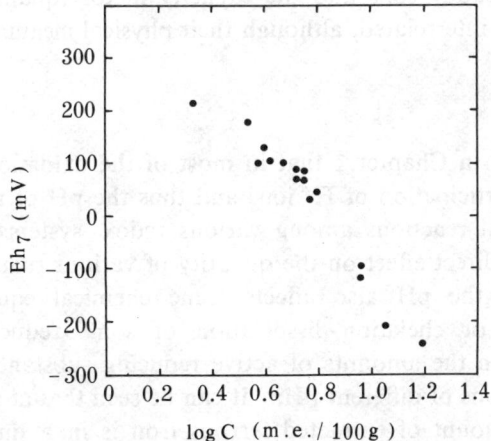

Fig. 2–9 Relationship between Eh_7 and amount of total reducing substances (chemical method)[6]

relation (Fig. 2-9), conforming to expectations deduced from Equation (2-3.) In another experiment the relationship between the logarithm of the amount of reducing substances determined by the electrochemical method (log C, where C is the diffusion current at $+0.7$ V) and the Eh is quite remarkable, with a correlation coefficient of -0.905 (Fig. 2-10).

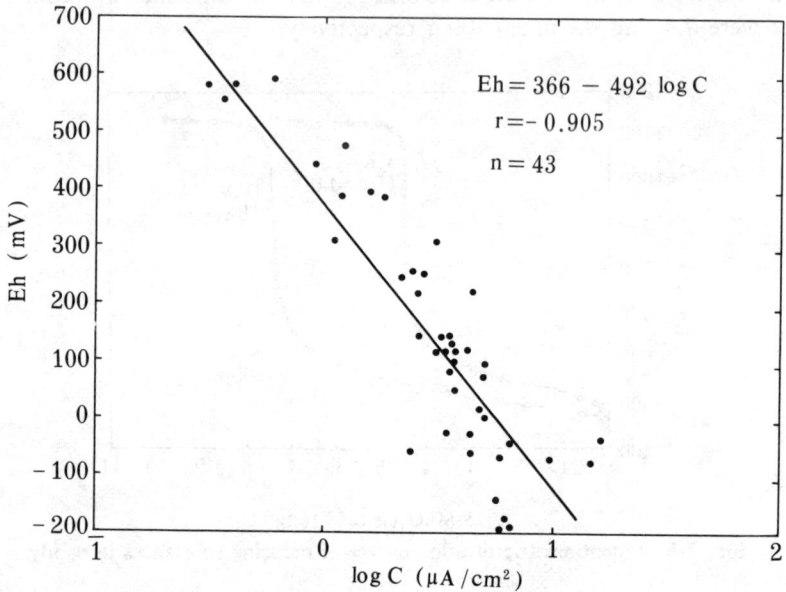

Fig. 2-10 Relationship between Eh and amount of reducing substances (electrochemical method)[8]

Thus, it may be concluded that the intensity factor *(Eh)* of oxidation-reduction property of paddy soils and the capacity factor (quantity of reducing substances) are closely interrelated, although their physical meanings are different.

2.2.3 pH

It has been stated in Chapter 1 that in most of the oxidation-reduction reactions there is the participation of H^+ ions and thus the pH of the medium can directly affect chemical reactions among various redox systems. As a result, the pH would have a direct effect on the quantity of various reducing substances of the soil. Besides, the pH also affects some chemical equilibria such as precipitation-solution or chelation-dissociation of some reducing substances. In Fig. 2-11 are shown the amounts of active reducing substances extracted by 0.1 M $Al_2(SO_4)_3$ solutions of different pH. It can be seen that at a pH of lower than about 3.3 the amount of extracted ferrous iron is in a linear correlation with the pH of the equilibrium solution. On the other hand, the amount of extracted active organic reducing substances only slightly increases when the

2.3 Dynamics of reducing substances in paddy soils

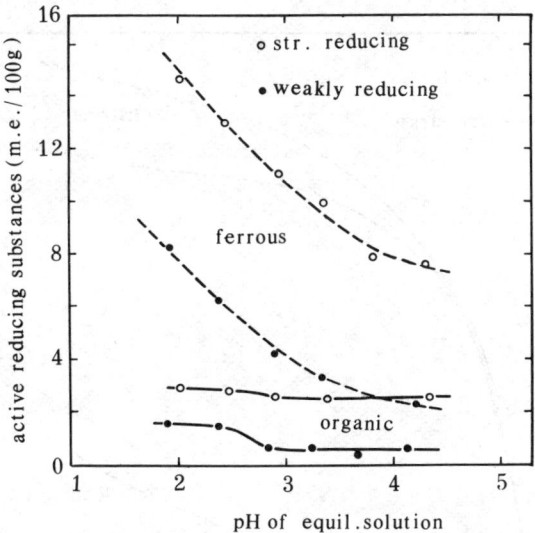

Fig. 2-11 Amount of active reducing substances as extracted at different pH of equilibrium solution[6]

pH of the equilibrium solution is lower than 2.5. Thus, it may be assumed that, contrary to reduced iron and manganese which are strongly pH-dependent with respect to their forms as will be discussed in more detail in Chapter 4, these organic substances are a group of substances easily extractable by water or solutions of neutral salts.

2.3 DYNAMICS OF REDUCING SUBSTANCES IN PADDY SOILS

2.3.1 Quantity

Following the progressive intensification of reduction after submerging, the amount of reducing substances in paddy soils increases. The rate of increase is dependent on the content of easily decomposable organic matter, temperature, and other soil factors which influence the decomposition of organic matter such as the pH. In Fig. 2-12 is shown the general pattern of change in the amount of active reducing substances after submerging for three paddy soils. The increase for a fertile neutral paddy soil derived from a lacustrine deposit was fairly rapid, whereas for an infertile acid paddy soil derived from yellow soil it was rather slow, and for a paddy soil derived from red soil, intermediate. After the intensive decomposition of organic matter the rate of increase lessened gradually, and the amount decreased again after a relatively stable period.

If a comparison is made among plant materials, it will be observed that the effect of vetch is greater than that of rice straw.

In Fig. 2-13 is shown an example measured in the field. In the cultivated layer the amount of reducing substances one month after submerging may be as high as 11—12 m.e./100 g. It decreased gradually afterwards. In plowpan

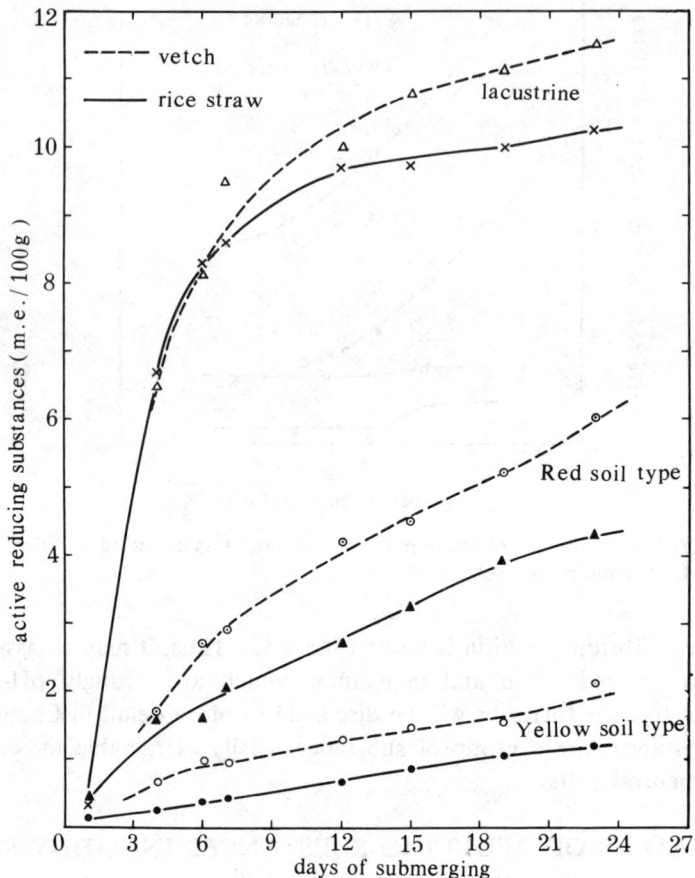

Fig. 2–12 Increase in amount of active reducing substances after submerging of paddy soil (plus 1% O.M.) (chemical method) (courtesy of S. H. Wu)

and the horizons underlying it the amount may also increase at the latter stages of rice growth due to the intensification of reduction condition, although to a much lesser extent than that in the cultivated layer (Fig. 2–13).

2.3.2 Changes in organic and inorganic components

In the course of the increase in the amount of reducing substances in paddy soils after submerging, the rates of increase for the organic component and the inorganic component are not the same. It can be seen from Figs. 2–14 and 2–15 that at the initial stage of submergence the active reducing substances produced are primarily organic in nature. The proportion of ferrous iron increases gradually at latter stages. And, after a certain period ferrous iron may account for the major part of active reducing substances. It is worthy of note that the

2.3 Dynamics of reducing substances in paddy soils

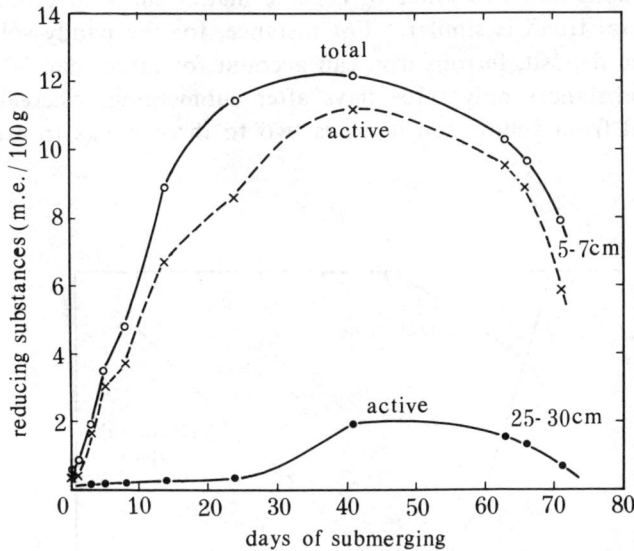

Fig. 2-13 Dynamics of reducing substances of the soil during the growing period of rice (chemical method)[1]

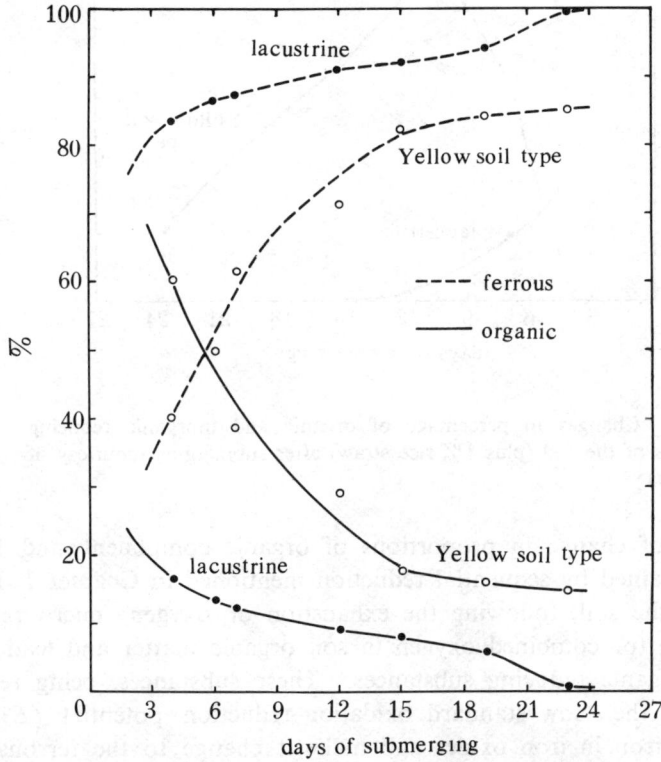

Fig. 2-14 Changes in percentage of organic and inorganic reducing substances of the soil (plus 1% vetch) after submerging (courtesy of S. H. Wu)

behaviors of two soils and two kinds of organic matter show some differences, although the general trend is similar. For instance, for the paddy soil derived from the lacustrine deposit, ferrous iron can account for more than 80% of the active reducing substances only three days after submerging, whereas for the paddy soil derived from yellow soil it needs two to three weeks to attain such a figure.

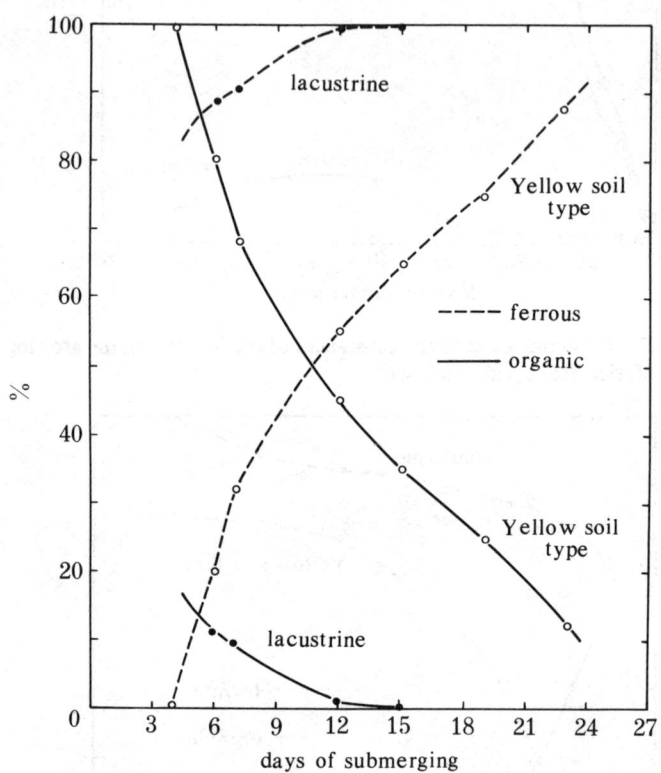

Fig. 2-15 Changes in percentage of organic and inorganic reducing substances of the soil (plus 1% rice straw) after submerging (courtesy of S. H. Wu)

This pattern of change in proportions of organic component and ferrous iron may be explained by sequential reduction mentioned in Chapter 1. During submergence of the soil, following the exhaustion of oxygen, microorganisms begin to compete for combined oxygen in soil organic matter and lead to the production of organic reducing substances. These substances, being reducing in nature due to their low standard oxidation-reduction potential (E^0), can react with ferric iron in iron oxides and make it change to the ferrous state. Therefore, it may be concluded that in the reduction process of soils the organic part of reducing substances is the most active and the determinative component, and the production of inorganic reducing substances such as ferrous iron is only

2.3 Dynamics of reducing substances in paddy soils

the result of further interactions between organic reducing substances and other electron acceptors in the soil.

2.3.3 Changes in fractions with different reduction intensity

In Fig. 2–16 is shown the change in reducing substances of different reduction intensity during submergence for a paddy soil with an added 3% of green manure. To summarize the results from a number of soils and incubation solutions of green manure, based on the difference in half–wave potential, the reducing substances may be classified into five groups, designated as (I)—(V) respectively. Group (I) appears only under strongly reducing conditions. Group (II) appears at the stage of intensive decomposition of organic matter and disappears afterwards. Group (III) is the most commonly encountered component, and disappears only at a high oxidation–reduction potential. Groups (IV) and (V) are also commonly found components.

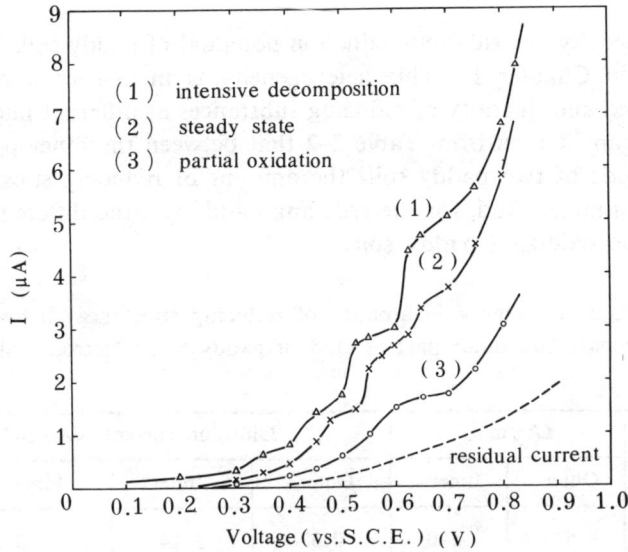

Fig. 2–16 Changes in reducing substances of different reduction intensity in paddy soil (plus 3% green manure) after submerging (electrochemical method)[3]

The decomposition products of green manure differ from those of the green manure added to soil in that in the former case groups (II) and (III) account for more than 70% of the total amount of the five groups, and in the latter case groups (IV) and (V) account for more than 55% of the total. The lower proportion of groups (II) and (III) in the latter case may probably be due to the oxidation of these components with lower standard potential by ferric and manganic oxides of the soil.

The reducing substances at the stage of intensive decomposition of organic matter are characterized by their large amount and complicated composition. At this stage the components with lower half–wave potential are also relatively large in proportion. They can usually be distinguished into 3—4 groups. With the progress in the decomposition process the amount of strongly reducing substances decreases gradually, accompanied by the decrease in the number of groups. At the steady stage of decomposition of organic matter there remain only one or two groups of reducing substances with high half–wave potential. If the soil is partially oxidized the amount of reducing substances decreases further.

The regularity of the change in reducing substances under different reduction intensity during submergence is in keeping with the general pattern of change in oxidation–reduction potential in paddy soils.

2.4 AMOUNT OF REDUCING SUBSTANCES IN PADDY SOILS

2.4.1 Heterogeneity in distribution of reducing substances in the soil

The heterogeneity in oxidation–reduction potential of paddy soils has already been discussed in Chapter 1. This heterogeneity is in essence a reflection of differences in kind and quantity of reducing substances at different micro–regions of the soil. It can be seen from Table 2-2 that between the inner part and the outer part of clods of two paddy soils the amount of reducing substances may differ by several times. And, for the reducing paddy soil the difference is larger than that for the oxidizing paddy soil.

TABLE 2-2 Difference in amount of reducing substances between the inner part and outer part of clod of paddy soil (electrochemical method)

Soil	Eh (mV)		Diffusion current ($\mu A / cm^2$)		
	Outer	Inner	Outer	Inner	Inner / Outer
Oxidizing	570	520	0.37	1.14	3.1
Reducing	580	330	0.73	4.04	5.5

The root system of rice is capable of secreting oxygen to the environment, thus causing some of the reducing substances in the root–zone to be oxidized. It can be seen from Table 2-3 that the amount of reducing substances in the root–zone is lower than that in the bulk soil for all of the three soils.

The heterogeneity in the distribution of reducing substances in paddy soils is a common phenomenon. According to statistics on 8—10 field measuring results for a number of soils, the mean variation coefficient is 26.5%, with 46.8% as the maximum and 10.1% as the minimum. This is the principal cause of the variation in Eh with different measurements in different localities as mentioned

2.4 Amount of reducing substances in paddy soils

in Chapter 1. This is also an important point which must be considered in dealing with the oxidation-reduction status of paddy soils.

TABLE 2-3 Difference in amount of reducing substances between bulk soil and root-zone of rice (electrochemical method)

Soil	Diffusion current ($\mu A / cm^2$)		
	Root-zone	Bulk soil	Bulk soil / Root-zone
Fertile paddy soil	3.40	5.07	1.5
Young paddy soil	2.40	4.96	2.1
Ordinary paddy soil	0.84	3.47	4.1

2.4.2 Distribution in the profile

In Fig. 2-17 is shown the distribution of reducing substances in profiles of three paddy soils and one upland soil.

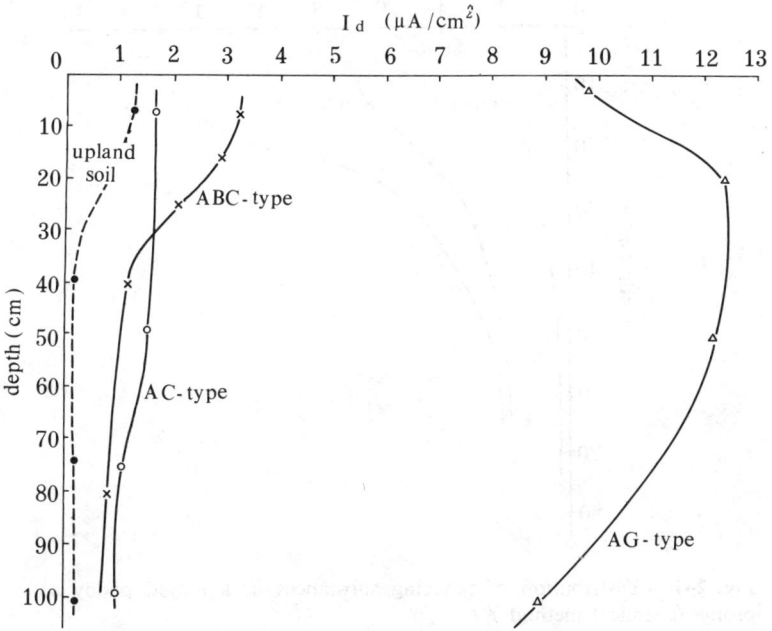

Fig. 2-17 Distribution of reducing substances in paddy soil profiles (electrochemical method)[8]

The A-C type soil is a young paddy soil derived from red soil. In the cultivated layer, the amount of reducing substances is rather low, with a diffusion current of 1.58 $\mu A / cm^2$. In the C horizon the current is 0.78 $\mu A / cm^2$, not very different from that in the upper layer.

The A–B–C type soil is an old paddy soil cultivated for rice for a long time. In the cultivated layer the organic matter content is quite high, and as a consequence the amount of reducing substances is relatively high, with a diffusion current of 3.22 $\mu A / cm^2$. The current decreases gradually from plowpan downwards, and attains a value of 0.77 $\mu A / cm^2$ at a depth of 80 cm.

The A–G type soil is a strongly reducing paddy soil. The content of reducing substances is high throughout the whole profile, with a diffusion current ranging from 9.85 to 13.3 $\mu A / cm^2$ within the one meter solum.

In the cultivated layer of upland soils there is also a small amount of reducing substances if the soil is sufficiently wet. However, in the C horizon the diffusion current is usually so small that it can hardly be distinguishable from the residual current.

In Fig. 2-18 is shown the distribution of various forms of reducing substances in the profile of a gleyed paddy soil. It can be seen that the general pattern is that the amounts of various forms of reducing substances decrease gradually from the cultivated layer downwards. The contents of ferrous iron, active reducing substances and total reducing substances between the surface layer and lower layers differ by a factor of 3, 5 and 6 respectively.

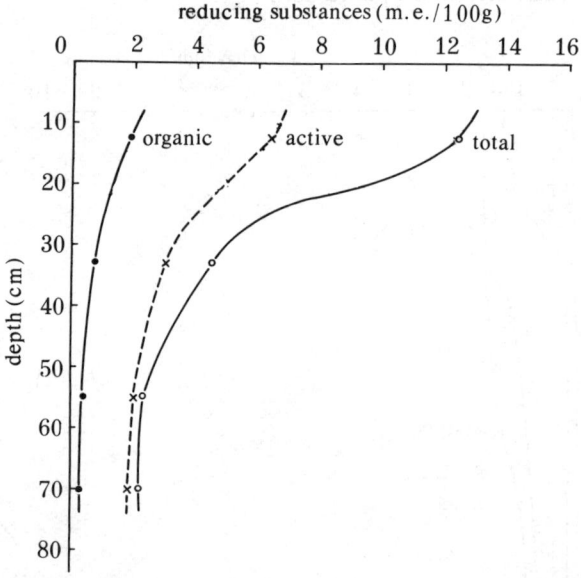

Fig. 2-18 Distribution of reducing substances in a gleyed paddy soil profile (chemical method)[7]

The heterogenous distribution of various reducing substances in the profile of paddy soils is of great significance in soil genesis.

2.4.3 Amount in paddy soils with different oxidation-reduction regime

In Table 2-4 is shown the content of reducing substances in paddy soils of different oxidation–reduction regimes. Some data for mud and upland soils

2.4 Amount of reducing susbtances in paddy soils

are also included for comparison. It can be seen that for the same great group of paddy soils the amount of reducing substances in different soil types may vary by a factor of ten or even twenty due to the difference in oxidation–reduction status of the soil, and the variation in amount of strongly reducing substances is greater than that of weakly reducing substances. It is also worthy of note that the stronger the reduction of the soil the larger the proportion of strongly reducing substances in the total content. These coincide with what would be expected from theoretical considerations.

TABLE 2–4 Amount of reducing substances in the surface layer of soils (milky stage of late rice) (electrochemical method)[8]

Soil type	Redox status	Eh (mV)	Reducing substances ($\mu A / cm^2$)			% in reducing substances	
			Strongly reducing	Weakly reducing	Total	Strongly reducing	Weakly reducing
Young paddy soil from red soil	Oxidizing	450	0.08	0.69	0.77	10.4	89.6
	Moderately reducing	110	1.83	10.7	12.5	14.6	85.4
	Strongly reducing	20	7.03	13.2	20.2	34.8	65.2
Paddy soil from red soil	Weakly reducing	400	0.39	2.72	3.11	12.5	87.5
	Moderately reducing	150	1.41	3.86	5.27	26.8	73.2
	Strongly reducing	–10	11.7	4.6	16.3	71.8	28.2
Paddy soil from alluvium	Oxidizing	560	0.06	2.15	2.21	2.7	97.3
	Weakly reducing	370	0.37	4.01	4.38	8.4	91.6
	Moderately reducing	150	4.47	4.87	9.34	47.9	52.1
Mud	Ordinary	310	1.23	3.50	4.73	26.0	74.0
	With rusty water	80	8.99	6.21	15.2	59.1	40.9
Upland soil	Oxidizing	450	0	1.16	1.16	0	100
Forest soil	Oxidizing	440	0	1.53	1.53	0	100

Based on the quantity of reducing substances as determined by the electrochemical method, i.e., the diffusion current at $+0.7$ V, the oxidation–reduction status of soils may be roughly classified into three types, namely the oxidizing, reducing and strongly reducing. The corresponding current is 0—1.5, 1.5—5.0 and 5.0—15 $\mu A / cm^2$ respectively. The Eh ranges of these soils are higher than 300, 300—50 and 50—(–200) mV respectively.

REFERENCES

(1) Institute of Soil Science, 1961. Soil Environment of High-yield Rice. Chapter 6. Science Press, Beijing.
(2) Yu Tian-ren et al., 1976. Electrochemical Properties of Soils and Their Research Methods (revised ed.). Chapter 11. Science Press, Beijing.
(3) Institute of Soil Science, 1978. Soils of China. Part B, Chapter 10. Science Press, Beijing.
(4) Yu Tian-ren, Zhang Xiao-nian et al., 1980. Electrochemical Methods and Their Applications in Soil Research. Chapter 11. Science Press, Beijing.
(5) Yu Tian-ren and Liu Wan-lan, 1957. Studies on oxidation-reduction processes in paddy soils. III. Effect of oxidation-reduction condition of the soil on rice growth. Acta Pedologica, **5**: 292-304.
(6) Liu Zhi-guang and Yu Tian-ren, 1962. Studies on oxidation-reduction processes in paddy soils. V. Determination of reducing substances. Acta Pedologica, **10**: 13-18.
(7) Lei Wen-jin and Zhu Hong-guan, 1964. Soils and their utilization in Lixiaho region, Jiangsu Province. Soils Bulletin, **36**: 130-178.
(8) Ding Chang-pu, Liu Zhi-guang and Yu Tian-ren, 1982. Determination of reducing substances in soils by voltammetric method. Soil Sci., **134**: 252-257.

CHAPTER 3

OXYGEN

PAN SHU-ZHENG

Oxygen participates in the periodical change of oxidation-reduction condition of paddy soils as a most active electron acceptor. When the oxygen content is high, the soil remains in an oxidizing state and the development of reduction processes is retarded. If the oxygen content is low, the soil will be reduced.

After the submerging of the soil, the direct supply of oxygen from the atmosphere is cut off. Under such circumstances oxygen consumption by microbiological oxidation-reduction reactions will be strengthened. As a result, a series of physico-chemical properties of the soil such as oxidation-reduction potential, pH and the form of some elements capable of participating in oxidation-reduction reactions (N, S, Fe, Mn, etc.) will be changed.

In this chapter the characteristics of the oxygen system are briefly dealt with, with special reference to the dynamics of oxygen and, finally, the contents of oxygen in representative paddy soils are presented.

3.1 CHARACTERISTICS OF THE OXYGEN SYSTEM

3.1.1 Oxygen as an oxidation-reduction system

The oxidation-reduction reaction of the oxygen system is:

$$O_2 + 4e + 4H^+ \rightleftarrows 2H_2O \tag{3-1}$$

According to the general equation for the oxidation-reduction reaction with the participation of H^+ ions:

$$pe = pe^0 + \frac{1}{n} \log \frac{(\text{oxidant})}{(\text{reductant})} - \frac{m}{n} pH \tag{3-2}$$

in the case of the oxygen system, we can write:

$$pe = pe^0 - pH + \frac{1}{4} \log (O_2) \tag{3-3}$$

$$Eh = E^0 - \frac{2.303 RT}{F} pH + \frac{2.303 RT}{4F} (O_2) \tag{3-4}$$

When the concentration of O_2 is 0.21 atmosphere:

$$pe = 20.8 - pH + \frac{1}{4} \log (0.21) \tag{3-5}$$

at 25°C

$$Eh = 1.23 - 0.059\, pH + 0.015 \log (0.21) \tag{3-6}$$

If the pH of the medium is 7, we get:
$$Eh = 810 \text{ mV}$$
$$pe = 13.7$$

Since the standard oxidation–reduction potential (E^0) of the oxygen system is the highest among the redox systems in paddy soils, oxygen is an oxidizing agent with a high degree of chemical reactivity. It can be calculated that in soils with a pH of 7 the oxidizing power of oxygen is higher than that of 0.1 M nitric acid when its partial pressure is as low as 10^{-4} atmosphere.

When oxygen reacts with reducing substances such as ferrous iron or glucose, the equations can be written as:

$$4Fe^{2+} + O_2 + 4H^+ = 4Fe^{3+} + 2H_2O \tag{3-7}$$
$$C_6H_{12}O_6 + 6O_2 = 6CO_2 + 6H_2O \tag{3-8}$$

3.1.2 Relationship between oxygen concentration and Eh

According to equation (3–6), the relationship between oxygen concentration and Eh at pH 7 is:

$$Eh = 0.82 + 0.015 \log (O_2) \tag{3-9}$$

Thus, from the theoretical viewpoint it is possible to calculate the Eh of the medium at different oxygen concentrations, if the oxidation–reduction potential of the medium is solely determined by oxygen. However, in practice, it has been observed that if the concentration of oxygen is determined with an O_2-sensor of the membrane–covered type, the Eh is measured with a platinum electrode and the pH is determined with a glass electrode, the difference between the calculated Eh value and the practically measured one may amount to several hundred millivolts (see Table 3–1).

TABLE 3-1 Comparison between measured and calculated Eh values for some soil solutions

Solution	pH	Eh (mV) Measured	Eh (mV) Calculated*
Buffer 1	6.82	378	815
Buffer 2	4.0	378	710
Filtrate from loamy paddy soil	8.1 8.4 8.5	240 227 223	740 698 687
Filtrate from acid paddy soil	4.5 4.6 4.6	402 405 401	705 709 706
Eluent from soil column	6.9 7.6 8.0	340 293 312	832 795 774

* Based on theoretical relationship between Eh and O_2–H_2O system

3.1 Characteristics of the oxygen system

In Figs. 3–1 and 3–2 are shown the results of an experiment conducted with two paddy soils. It can be seen that although there is a certain correlation between the Eh and $\log(O_2)$, the position of points on the figure is rather scattered, and the slope is much steeper than that of the theoretical Eh_7–$\log(O_2)$ line.

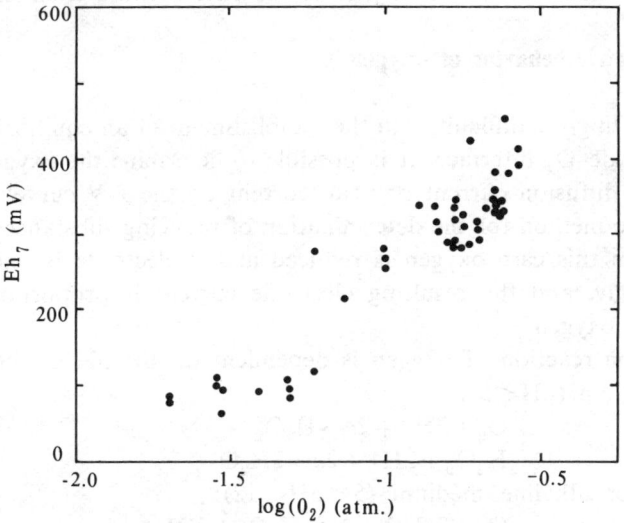

Fig. 3–1 Relationship between Eh_7 and $\log(O_2)$ for a sandy paddy soil

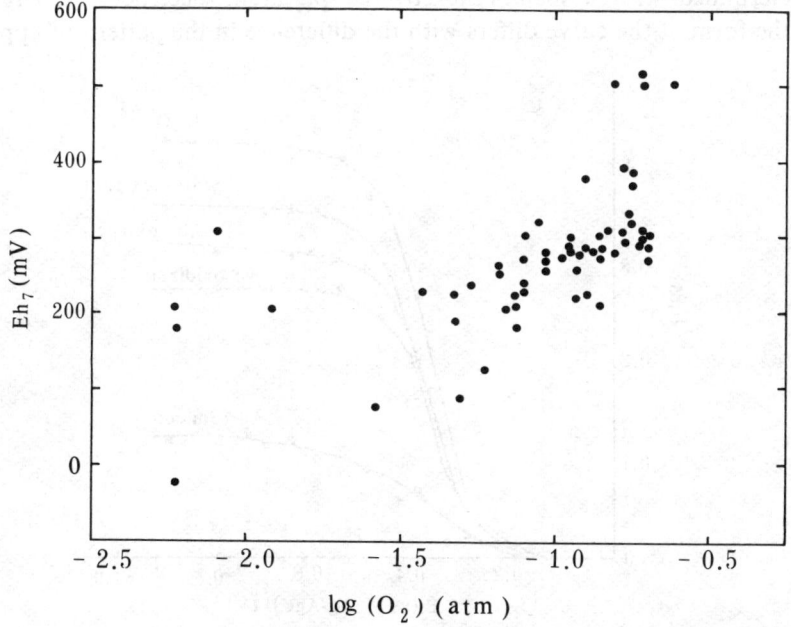

Fig. 3–2 Relationship between Eh_7 and $\log(O_2)$ for a paddy soil derived from red soil

The discrepancy is assumed to be caused primarily by the formation of an oxide film on the surface of the platinum electrode in the presence of oxygen (cf. Chapter 1). In such case the platinum electrode will not function as a reversible oxidation–reduction electrode. Therefore, in complicated soil systems it is not possible to directly calculate the concentration of oxygen from the *Eh* value measured by conventional methods, just as the case in pure chemical systems.

3.1.3 Voltammetric behavior of oxygen

Notwithstanding the difficulties in the establishment of an equilibrium potential at the electrode–O_2 interface, it is possible to determine the oxygen concentration from the diffusion current or peak current on the I–V curve[3], similar in principle to the method for the determination of reducing substances described in Chapter 2. In this case oxygen is reduced at the electrode by electrons introduced externally, and the resulting electrode current is proportional to the concentration of oxygen.

The reduction reaction of oxygen is dependent on the pH of the medium:
In acid medium (pH<5):
$$O_2+2H^++2e \rightarrow H_2O_2 \qquad (3-10)$$
$$H_2O_2+2H^++2e \rightarrow 2H_2O \qquad (3-11)$$
In neutral or alkaline medium (5<pH<12):
$$O_2+2H_2O+2e \rightarrow H_2O_2+2OH^- \qquad (3-12)$$
$$H_2O_2+2e \rightarrow 2OH^- \qquad (3-13)$$

In Figs. 3–3, 3–4 and 3–5 are shown I–V curves of dissolved oxygen in the soil as determined with a membrane-covered platinum electrode. It can be seen that the form of the curve differs with the difference in the pattern of applied

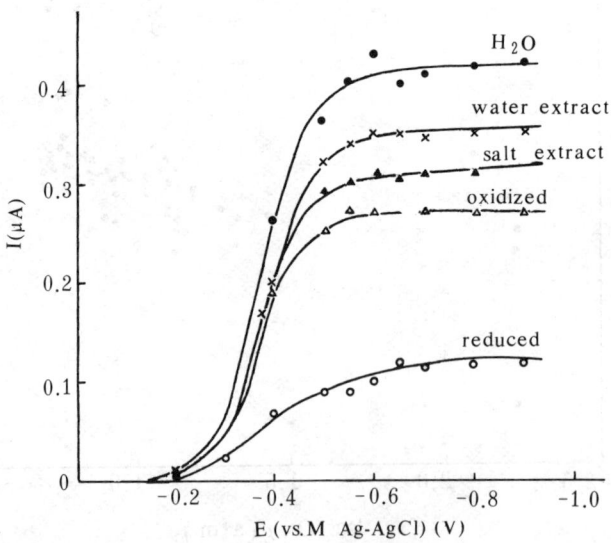

Fig. 3–3 I–V curves for dissolved oxygen in soils (classical method)[3]

3.1 Characteristics of the oxygen system

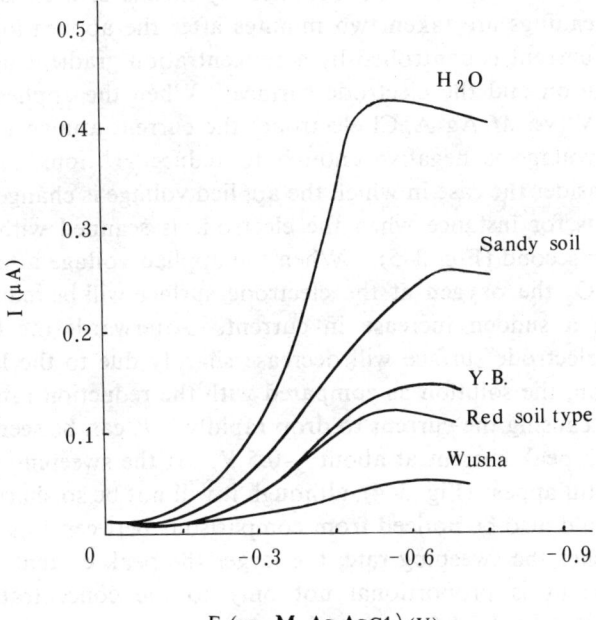

Fig. 3–4 I–V curves for dissolved oxygen in soils (sweeping rate 5 mV/sec)

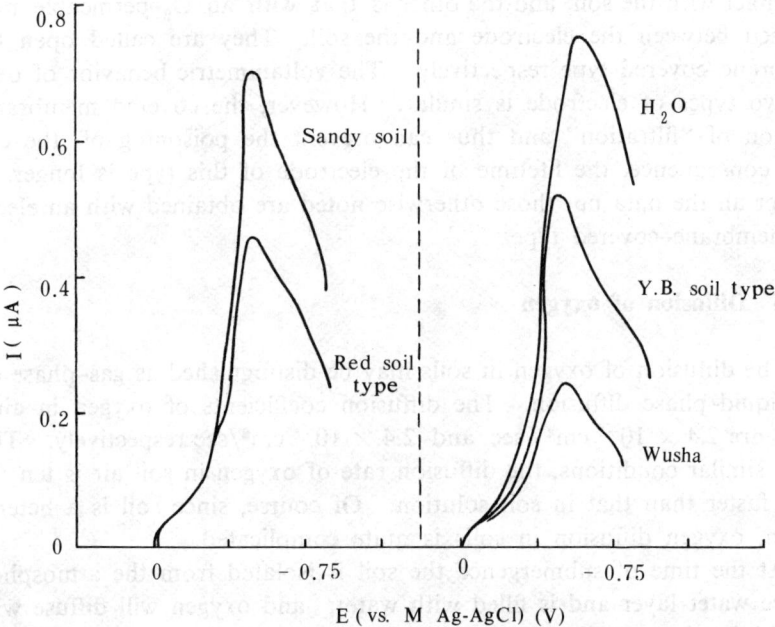

Fig. 3–5 I–V curves for dissolved oxygen in soils (sweeping rate 50 mV/sec)

voltage. Curves of Fig. 3–3 are obtained by means of a classical method, and the current readings are taken two minutes after the application of voltage. In this case the current is controlled by a concentration gradient of oxygen between the bulk solution and the electrode surface. When the applied voltage exceeds about -0.6 V (vs. M Ag–AgCl electrode) the current attains a steady value until the applied voltage is negative enough to reduce H^+ ions.

Now consider the case in which the applied voltage is changed at a sufficiently high speed, as for instance when the electrode is scanned with a sweeping rate of 50 mV per second (Fig. 3–5). When the applied voltage attains the reduction potential of O_2 the oxygen at the electrode surface will be reduced immediately, thus causing a sudden increase in current. Afterwards the O_2 concentration close to the electrode surface will decrease sharply due to the low diffusion rate of oxygen from the solution as compared with the reduction rate at the electrode surface, thus causing the current to drop rapidly. It can be seen from the figure that there is a peak current at about -0.5 V. If the sweeping rate is low a peak current will still appear (Fig. 3–4), although it will not be so sharp as in the former case. It should also be noticed from comparisons between Fig. 3–4 and Fig. 3–5 that the quicker the sweeping rate, the larger the peak current. This is because the peak current is proportional not only to the concentration of dissolved oxygen, but also to the 1/2 power of sweeping rate.

The voltammetric behavior of oxygen mentioned above is the theoretical basis for the determination of oxygen with the electrochemical method. In practical work, the most commonly used electrode is a platinum electrode. There are two types of such electrode (sensor). One type is the electrode which is directly in contact with the soil, and the other is that with an O_2–permeable membrane imposed between the electrode and the soil. They are called open type and membrane–covered type respectively. The voltammetric behavior of oxygen on the two types of electrode is similar. However, the covered membrane has a function of "filtration" and thus can prevent the poisoning of the electrode. As a consequence, the lifetime of the electrode of this type is longer. In this chapter all the data but those otherwise noted are obtained with an electrode of the membrane–covered type.

3.1.4 Diffusion of oxygen

The diffusion of oxygen in soils may be distinguished as gas–phase diffusion and liquid–phase diffusion. The diffusion coefficients of oxygen in air and in water are 2.4×10^{-1} cm^2/sec and 2.4×10^{-5} cm^2/sec respectively. Therefore, under similar conditions, the diffusion rate of oxygen in soil air is ten thousand times faster than that in soil solution. Of course, since soil is a heterogenous system, oxygen diffusion in soils is quite complicated.

At the time of submergence the soil is isolated from the atmosphere by a surface water layer and is filled with water, and oxygen will diffuse within the soil chiefly through the liquid phase. In addition to the concentration gradient of oxygen, the diffusion is also affected by the obstruction of soil particles. The effective diffusion coefficient (D_e) is[4]:

$$D_e = D_0 r \gamma s \tag{3-14}$$

where D_0 is the diffusion coefficient of oxygen in the liquid phase, r a tortuosity factor in the diffusion path, γ a factor relating to interactions between oxygen and soil particles, and s is the volume fraction not occupied by soil particles. D_0 is temperature-dependent, the temperature coefficient of it being about 4% at room temperature. The numeral value of γ for neutral oxygen molecules is close to 1.0. The coefficient r is used for a correction of the geometrical factor. For, in a heterogeneous medium like soil consisting of both solid phase and liquid phase, the dissolved oxygen can diffuse only through solution surrounding soil particles, and thus the actual diffusion distance would be much longer than that of the diffusion through a straight line.

From the above it can be well understood that oxygen diffusion will be affected by water content, texture and structure of the soil.

When soil is unsaturated with water there will be both a liquid-phase diffusion and a gas-phase diffusion of oxygen. Under such circumstances the oxygen diffusion rate will be much larger than that in water-saturated soils.

3.2 SOURCE AND CONSUMPTION OF OXYGEN IN PADDY SOILS

Submergence of the soil causes the cut-off of the direct supply of oxygen from the atmosphere, and promotes the biological consumption of oxygen. Therefore, there is a contradiction between the high rate of oxygen consumption and the low rate of oxygen supply. This is why the cultivated layer of paddy soil is generally deficient in oxygen under submerged conditions. In the following, the dynamic balance of oxygen in paddy soils will be discussed.

3.2.1 Source of oxygen

The oxygen in submerged soils comes from the atmosphere, rain-water, irrigation water and secretions from roots of rice and some other aquatic plants.

In Table 3-2 is listed the oxygen content of some natural waters. The oxygen content of ordinary pond water and ditch water is about 8.3 mg / liter at 20°C, corresponding to about 90% of that saturated with atmospheric oxygen. In some ponds growing with water hyacinth the oxygen content is 6.9 mg / liter. At a temperature of 14°C the oxygen content of pond and ditch waters is 9.5 mg / liter on the average due to the increased solubility of oxygen. It is conceivable that during the summer season with a temperature of 30°C or higher the oxygen content of natural waters would be lower than that listed in Table 3-2.

There is evidence showing that the oxygen content of natural waters differs with the flowing path from their sources. For example, it was observed that the oxygen index of a spring water determined at the mouth of the spring with a micro-platinum electrode was 2.45 μA, whereas it increased to 5.76 μA at a certain distance from the mouth, due apparently to the dissolving of atmospheric oxygen during the flow of water. Another example showed that the oxygen index of the water in an irrigation canal was 7.32 μA, and for surface waters in

TABLE 3–2 Oxygen content of some natural waters

Temperature (°C)	Source	O_2 (mg/l)
20	Saturated with O_2	9.2
	Pond 1	6.9
	Pond 2	8.9
	Pond 3	8.1
	Pond 4	6.9
	Pond 5	7.7
	Ditch 1	8.8
	Ditch 2	8.0
	River	8.5
	Waste water from factory	8.3
14	Saturated with O_2	10.4
	Pond 6	9.3
	Pond 7	9.1
	Pond 8	9.8
	Pond 9	9.6
	Pond 10	8.9
	Ditch 3	10.4

the paddy field 1.5 meter from the canal and for that in the center of the field it decreased to 5.96 and 3.89 μA respectively[2].

3.2.2 Consumption of oxygen

The oxygen in paddy soils is consumed through the following ways: (a) to function as an electron acceptor in microbiological respiration; (b) to be consumed in biological oxidation of NH_4^+ to NO_3^- or S^{2-} and S to SO_4^{2-}, etc.; and (c) to be used in chemical oxidations of Fe^{2+}, Mn^{2+}, S^{2-} and organic reducing substances. Among these processes (a) and (b) are relatively slow in speed, whereas (c) is more rapid. It is not the purpose of this chapter to deal with the biological processes under items (a) and (b), and in the following only the role of some reducing substances in the chemical consumption of oxygen will be discussed.

3.2.2.1 *Ferrous ions*

The chemical reaction between ferrous ions and oxygen in soils proceeds at a fairly rapid speed. In Fig. 3-6 are shown the results of such an experiment conducted at room temperature. In the treatment in which a low amount of Fe^{2+} was added the oxygen content decreased by 2.8 mg per liter 3 hours after the addition, corresponding to 77.8 mg of oxygen per mole of Fe^{2+}. For the medium and high amount treatments the decreases were 7.0 and 9.8 mg per liter respectively, corresponding to 24.6 and 17.2 mg of oxygen per mole of Fe^{2+} respectively.

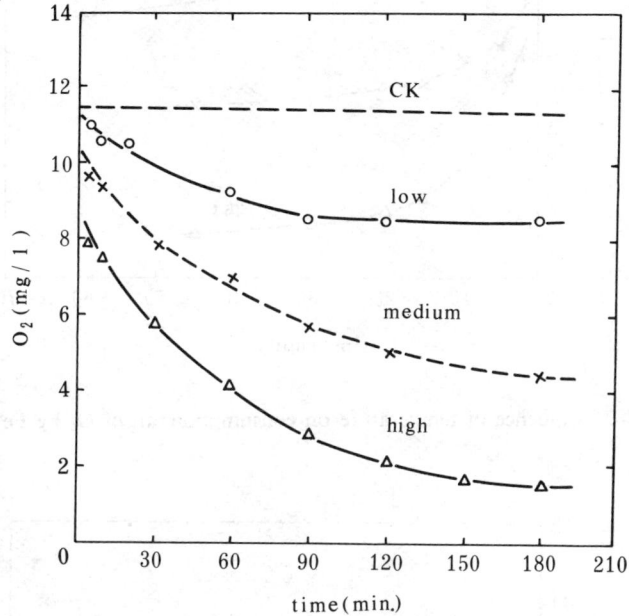

Fig. 3-6 Change in O_2 content of a paddy soil after the addition of Fe^{2+} (low, medium and high correspond to 1, 8 and 16 g $FeSO_4$ per 100 g soil respectively)

Temperature can influence the reaction rate between oxygen and ferrous iron. It can be seen from Fig. 3-7 that when a medium amount of Fe^{2+} was added the higher the temperature the quicker was the rate of decrease in oxygen content. After 30 minutes' reaction the oxygen content in the 35°C treatment was only 18% of that in the 19°C treatment, and in the latter case the soil still contained 1.8 mg per liter of oxygen even after a reaction time of 5 hours.

3.2.2.2 *Manganous ions*

The effect of manganous ions on oxygen is similar to that of ferrous iron. However, because of the much higher standard oxidation-reduction potential of the manganese system than that of the iron system, the speed of oxidation of Mn^{2+} is lower than that of Fe^{2+} (Fig. 3-8). For instance, when 16 g of $FeSO_4 \cdot 7H_2O$ (equivalent to $0.57M$ Fe^{2+}) was added the oxygen content after 60 minutes'

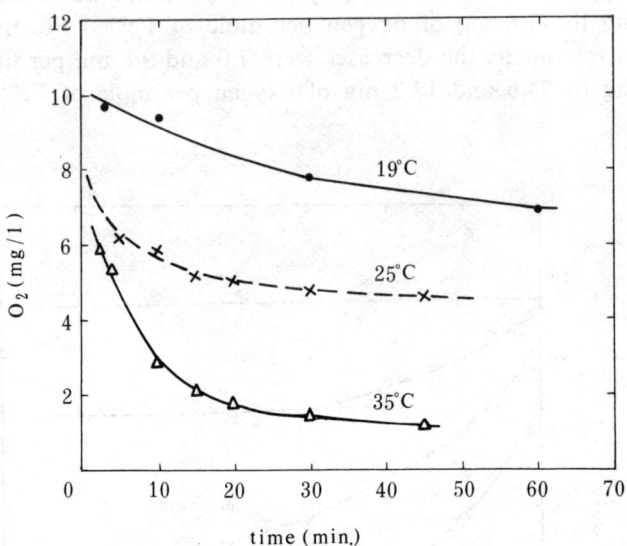

Fig. 3-7 Influence of temperature on consumption rate of O_2 by Fe^{2+}

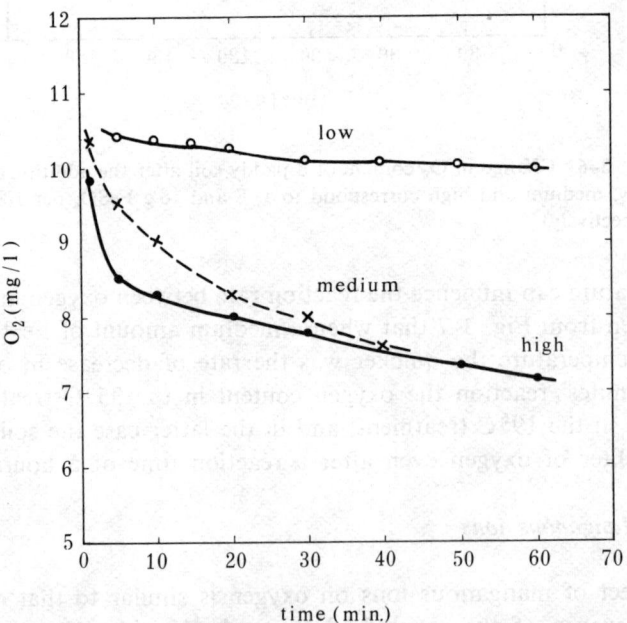

Fig. 3-8 Change in O_2 content of a paddy soil after the addition of Mn^{2+} (low, medium and high correspond to 8, 16 and 24 g $MnCl_2$ per 100 g soil respectively)

3.2 Source and consumption of oxygen in paddy soils

reaction decreased from 11.3 mg/l to 4.1 mg/l, whereas when 16 g of $MnCl_2 \cdot 4H_2O$ (equivalent to $0.8 M$ Mn^{2+}) was added the corresponding figures were 10.4 and 7.1 mg/l respectively. If calculated on a molar basis of reductant, the consumption of oxygen by Fe^{2+} was 12.6 mg/l, and that by Mn^{2+} was 4.1 mg/l.

3.2.2.3 Sulfide ions

Sulfide ions formed in submerged soils functioning as a kind of strong reducing agent can react with oxygen quickly. Fig. 3-9 shows that the oxygen content of the soil decreased rapidly after the addition of S^{2-}, especially if a high amount of S^{2-} was added.

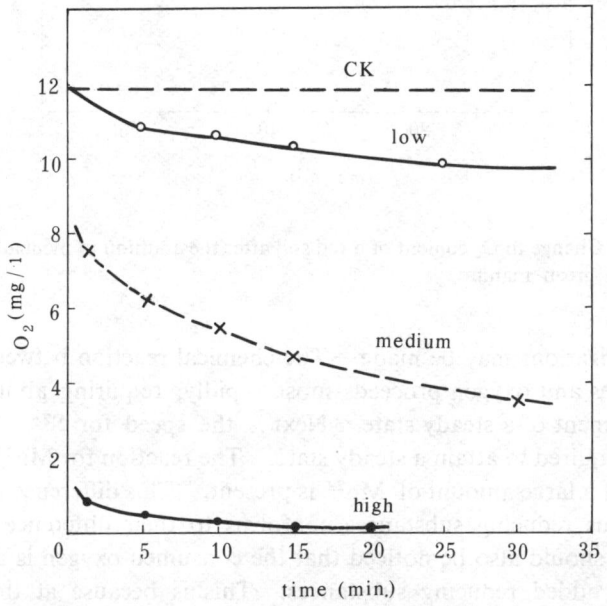

Fig. 3-9 Change in O_2 content of a paddy soil after the addition of S^{2-} (low, medium and high correspond to 0.08, 0.2 and 0.4 g Na_2S per 100 g soil respectively)

3.2.2.4 Organic reducing substances

In addition to inorganic substances such as Fe^{2+}, Mn^{2+} and S^{2-} mentioned above, organic reducing substances can also react with oxygen. Since the standard oxidation–reduction potential of these substances is low, the reaction with oxygen proceeds rapidly (Fig. 3-10). For instance, it was observed in an experiment that the oxygen content dropped to 4.2 mg/l within 1—2 minutes after the addition of a high amount of incubation solution of milk vetch.

To summarize the experimental results shown in Figs. 3-6 to 3-10, some

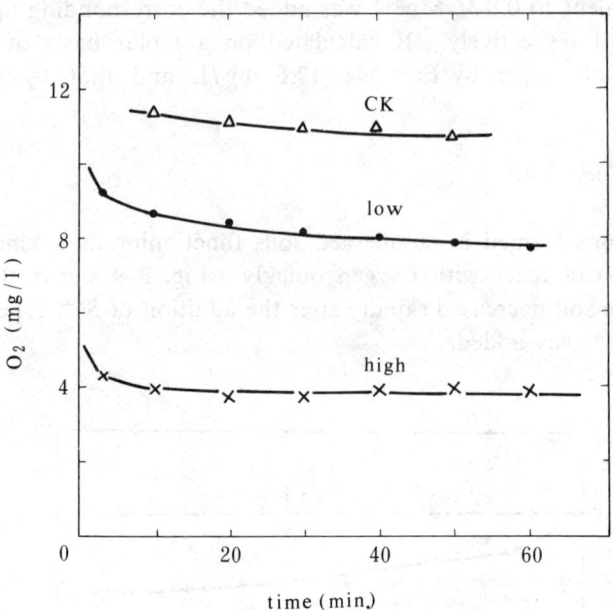

Fig. 3-10 Change in O_2 content of a red soil after the addition of incubation solution of green manure

qualitative generalizations may be made. The chemical reaction between organic reducing substances and oxygen proceeds most rapidly, requiring about 20 minutes for the attainment of a steady state. Next is the speed for S^{2-}. For Fe^{2+} several hours is required to attain a steady state. The reaction for Mn^{2+} proceeds most slowly even if a large amount of Mn^{2+} is present. This difference in reaction rate among various reducing substances conforms to their difference in reduction intensity. It should also be noticed that the consumed oxygen is not stoichiometrical to the added reducing substances. This is because at the time of determination of oxygen a true chemical equilibrium between oxygen and the reducing substances has not been attained. Besides, other soil factors may affect the reaction. These results imply that oxygen as an oxidation-reduction system is not reversible thermodynamically in soils, just as the case in pure solutions.

3.2.3 Dynamic balance of oxygen

Summarizing our knowledge of the source and consumption of oxygen, the dynamic balance of oxygen in paddy soils may be diagrammed in Fig. 3-11. Following the submerging of a soil the role of various oxygen-consuming factors is intensified, and eventually the oxygen content declines to a constant and low value when the soil attains a steady state.

The rate of decrease in oxygen content in soils after submerging is determined

3.2 Source and consumption of oxygen in paddy soils

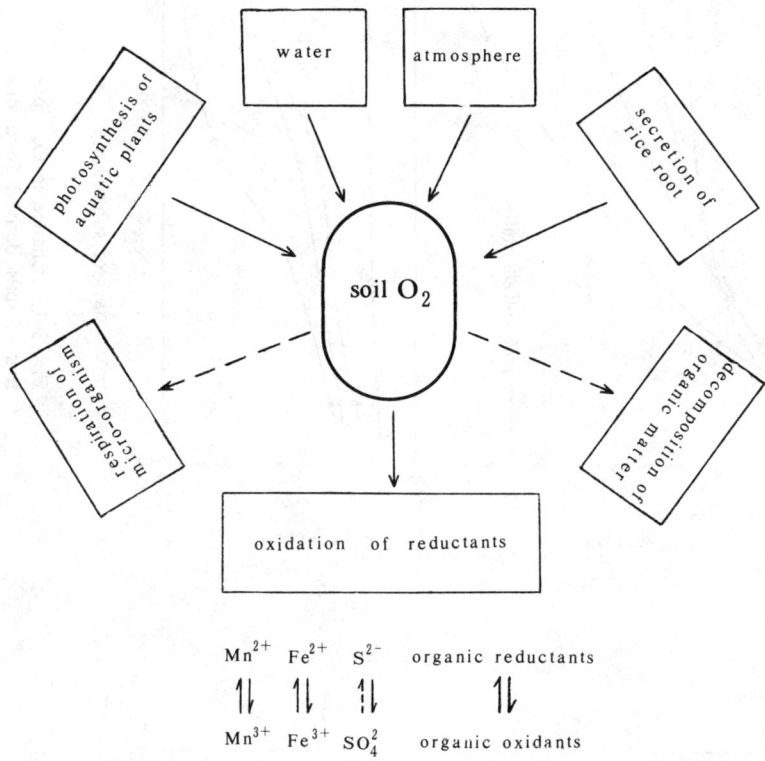

Fig. 3–11 Schematic diagram of source and consumption of oxygen in soil

by the rate of oxygen consumption. As far as soil factors are concerned the organic matter status and other factors affecting the biological decomposition of organic matter would play a dominant role. It can be seen from the experimental results shown in Fig. 3–12 that for the subsoil of a red soil with a very low organic matter content and a low pH there was no detectable consumption of oxygen during submergence, whereas if fresh organic matter was added the oxygen content would decline to a very low level only several hours after submerging during the summer season, and it practically disappeared one day after. Similarly, for the surface soil the rate of oxygen consumption was also high. If a comparison is made between the oxygen content curve and the Eh curve it should be noticed that the consumption of oxygen was closely related to the production of reducing substances. For different soils the rate and extent of decline in oxygen content differed with soil conditions. For instance, the decrease for an acid sulfate soil of pH 2 was slower than another acid sulfate soil with a pH of 4 (Fig. 3–13), due presumably to the retardation of microbiological activities by the strong acidity. This latter assumption is supported by the fact that the Eh of the soil practically did not change within one week after submerging. The fertility level

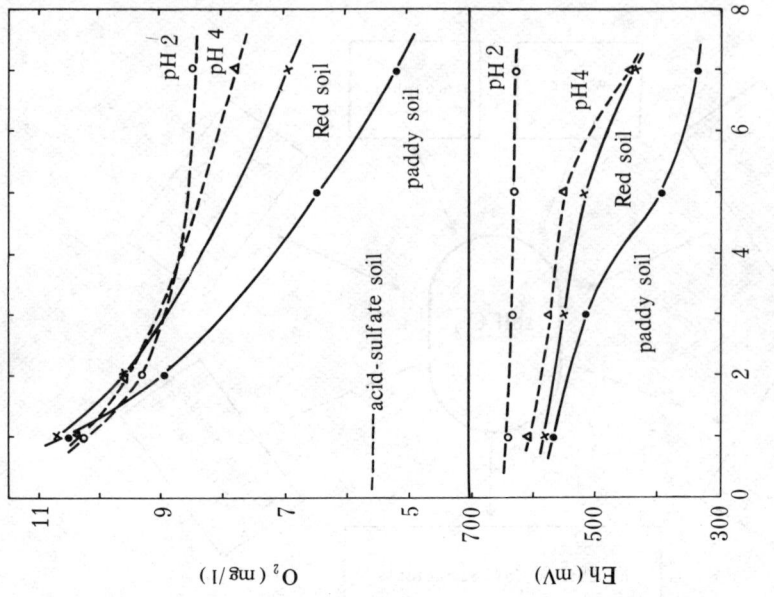

Fig. 3-13 Change in O$_2$ content of soils derived from different parent materials after submerging (room temperature)

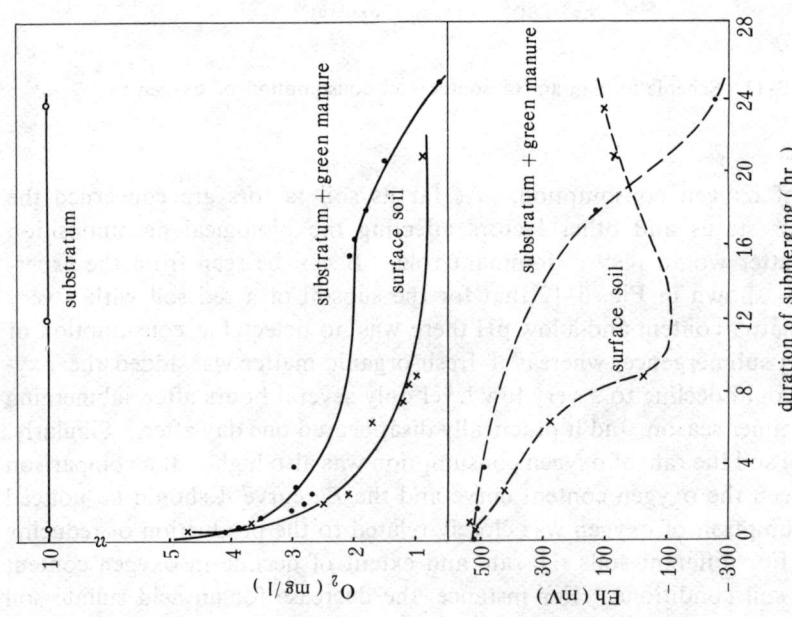

Fig. 3-12 Effect of organic matter on O$_2$ consumption in submerged soil (summer)

3.2 Source and consumption of oxygen in paddy soils

of a paddy soil derived from red soil is relatively high, hence the rate of decline in oxygen content was high.

In Fig. 3-14 is shown the change in oxygen content after submerging for four paddy soils. For the fertile soil 4 with a high content of organic matter (3.3%) the oxygen was almost exhausted at the fourth day after submerging. In soil 3 with 2% of organic matter the oxygen content was 6.6 mg/l after 7 days' submergence. For soil 1 with a low organic matter content and a pH of 5 the oxygen consumption was the least, due to weak microbiological activities.

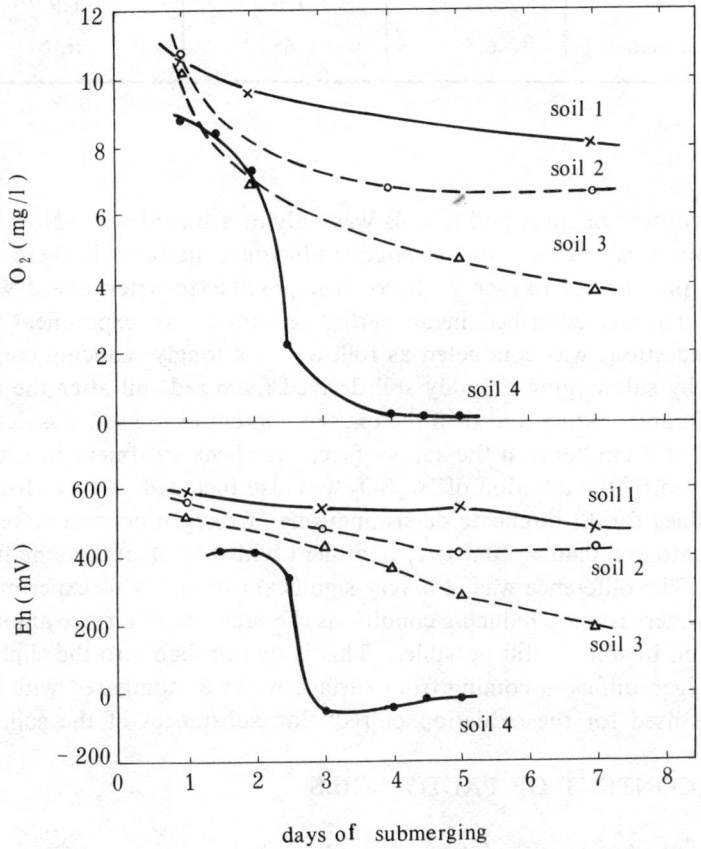

Fig. 3-14 Change in O_2 content of paddy soils (room temperature)

When the balance between supply and consumption of oxygen attains a steady state at the latter period of submergence the oxygen content of the soil is dependent on the relative rates of supply and consumption. In Table 3-3 is shown the oxygen content of some soils after two months' submergence. For the three paddy soils the oxygen was exhausted almost completely, whereas for the three upland soils there was still a certain amount of oxygen even after a long period of submergence, especially for the laterite with a low pH.

TABLE 3-3 Oxygen content of some submerged soils

Soil	pH	Eh (mV)	O_2 (mg/l)
Yellow–brown soil	7.2	323	1.6
Paddy soil from above	7.3	32	tr.
Red soil	6.0	304	0.8
Paddy soil from above	7.4	14	tr.
Laterite	4.9	120	1.9
Paddy soil from above	6.5	65	tr.

The oxygen content of three paddy soils was only at a trace level. Now the following questions arise: Under still stronger reducing conditions is there the possibility of the presence of oxygen? If so, is it possible to determine it with the voltammetric method described in an earlier section? An experiment for answering these questions was conducted as follows: A strongly reducing condition was created by submerging a paddy soil derived from red soil after the addition of 1% of organic matter, and then the oxygen content of the soil was determined at a depth of 3 cm beneath the soil surface. A check treatment in which oxygen was removed by the addition of Na_2SO_3 was also included. It was found that the mean values for 12 duplicate determinations of oxygen content were as follows: check treatment, 0.46×10^{-3} μA; treatment with 1% of organic matter, $1.18 \times 10^{-3} \mu A$. The difference was at a very significant level. This experiment shows that even under strongly reducing conditions the presence of a trace amount of dissolved oxygen in soils is still possible. This is due probably to the slightly higher rate of oxygen diffusion coming from surface water as compared with the consumption rate used for the oxidation of reducing substances of the soil.

3.3 OXYGEN CONTENT OF PADDY SOILS

3.3.1 Heterogeneity in oxygen content of paddy soils

At different localities of the cultivated layer of paddy soils there is a micro–regional difference in oxygen content due to differences in soil structure and organic matter status etc. It can be seen from Table 3-4 that for a given paddy field the difference in oxygen content among five measuring points may range by as high as 60%.

The root system of the rice plant is capable of secreting oxygen to the soil, resulting in the higher oxygen content of the root–zone, as demonstrated in Table 3-5.

The micro–regional variation in oxygen content in paddy soils is a common

3.3 Oxygen content of paddy soils

TABLE 3-4 Local variation of O_2 content in paddy fields (depth 3-4 cm) (June) (Jiangxi)

Measuring point	O_2 (mg/l)			
	Field 1	Field 2	Field 3	Field 4
1	1.6	2.0	1.5	1.9
2	1.4	1.6	1.7	1.7
3	1.7	1.6	1.6	1.9
4	1.8	1.2	1.4	1.9
5	2.0	1.7	1.2	2.0
Mean	1.7	1.6	1.5	1.9

TABLE 3-5 Oxygen contents of root-rich and root-poor regions of paddy field (earing stage of rice) (Nanjing)

Depth (cm)	O_2 (mg/l)			
	Root-poor (compact)	Root-poor (soft)	Root-rich	Root-rich (soft)
1	0.05	1.7	3.0	4.5
2	tr.	—	1.8	2.8
3	tr.	—	0.77	1.8
4	tr.	0.7	0.3	0.6
5	tr.	—	0.12	0.05
6	tr.	0.1	0.04	tr.
10	tr.	0.04	0.03	—

phenomenon. This is the basic cause of micro-regional differences in oxidation-reduction potential (Chapter 1) and amount of reducing substances (Chapter 2).

3.3.2 Distribution in the surface layer

In submerged soils with a surface water layer the oxygen may be supplied through the downward movement of percolating water and by the diffusion from the surface water, although the direct supply from atmospheric oxygen is cut off. Obviously, the shorter the distance from the water layer, the higher the oxygen content. In Fig. 3-15 is shown the distribution of oxygen in a submerged soil measured in the laboratory.

In Figs. 3-16, 3-17 and 3-18 is shown the gradual decrease in oxygen content with depth in the cultivated layer for three types of paddy soil. The curves in each figure are the results for different paddy fields of the same soil type. It can be seen that there usually exists a certain amount of dissolved oxygen even at a depth of several centimeters from the surface. It will also be noticed that at a depth of 1—2 cm the oxygen content for the paddy field developed on red

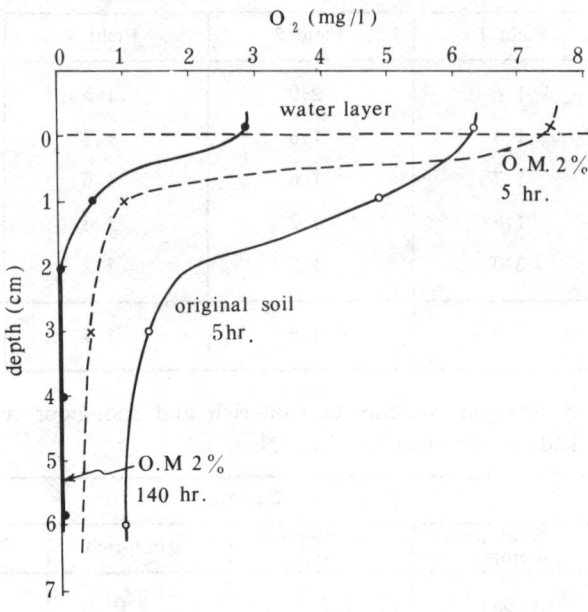

Fig. 3-15 O_2 content in different depths of submerged paddy soil

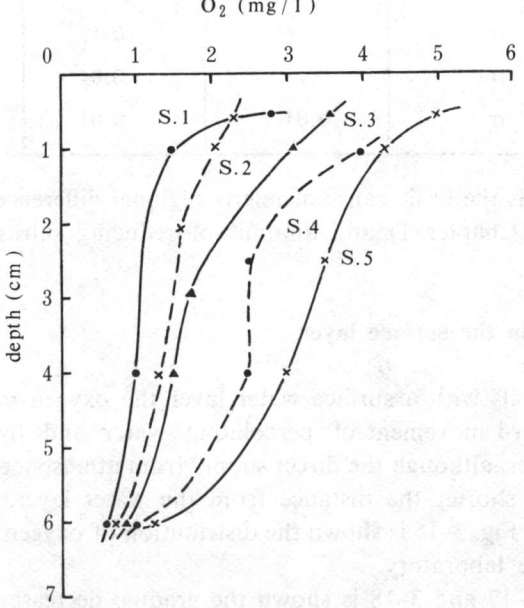

Fig. 3-16 O_2 content of the surface layer for submerged paddy soils derived from red soil (Jiangxi, June)

3.3 Oxygen content of paddy soils

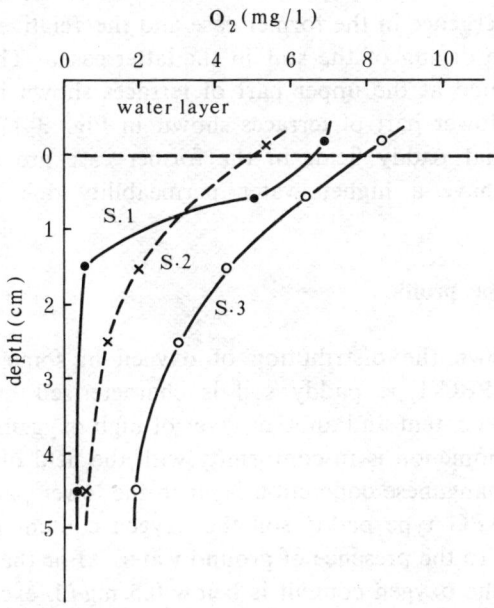

Fig. 3-17 O$_2$ content of the surface layer for submerged paddy soils derived from yellow-brown soil located at the lower part of terraces (Nanjing, October)

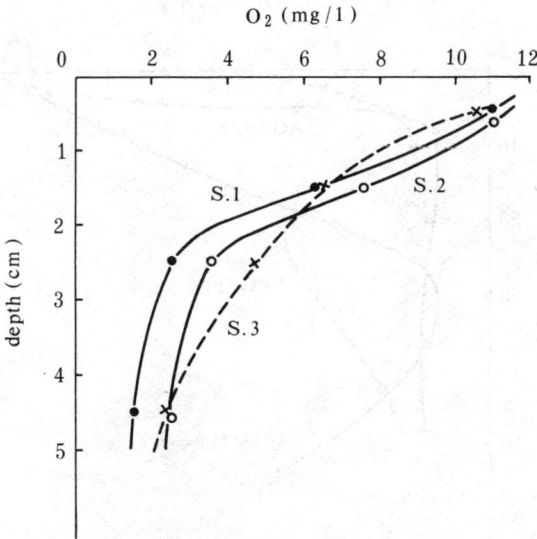

Fig. 3-18 O$_2$ content of the surface layer for submerged paddy soils derived from yellow-brown soil located at the upper part of terraces (Nanjing, October)

soil is lower than that for a paddy field developed on yellow-brown soil. This is presumed to be due to the high temperature at the time of measurement (July) and the long-term submergence in the former case and the relatively low temperature (October) and the drying of the soil in the latter case. The oxygen content in paddy fields located at the upper part of terraces shown in Fig. 3-18 is higher than that at the lower part of terraces shown in Fig. 3-17. The causes of this difference are that paddy fields in the former case are less influenced by ground water and have a higher water permeability due to lighter soil texture.

3.3.3 Distribution in the profile

In Fig. 3-19 is shown the distribution of oxygen in some representative paddy profiles. The APBC-type paddy soil is characterized by a cultivated layer with a high oxygen content and another layer of high oxygen content below 40 cm. This latter phenomenon is in conformity with the field observation that there is a distinct iron-manganese concretion layer in the lower part of the profile. In the profile of APG-type paddy soil the oxygen content below a depth of 30 cm is very low due to the presence of ground water. For the whole profile of AG-type paddy soil the oxygen content is below 0.5 mg/l, except in the cultivated layer which has been dried for some time.

The difference in oxygen content among different types of paddy soil is of great significance in soil genesis. This will be discussed in Chapter 9.

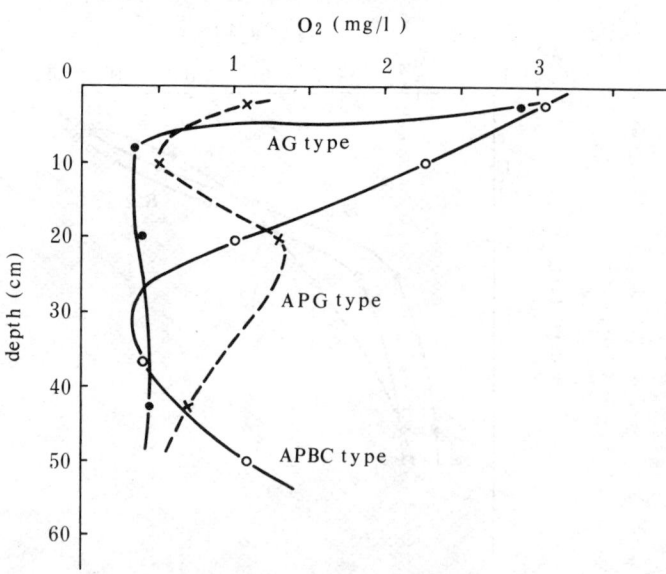

Fig. 3-19 O_2 content of some paddy soil profiles with different water regimes (Nanjing, October) (measured with micro-Pt electrode)

3.3.4 Effect of water management

Agricultural measures such as water management may have a profound influence on the oxygen content of the soil. In Fig. 3–20 are shown the different effects of two kinds of irrigation on oxygen content of soil. It is clear that the oxygen content in the cultivated layer of the wet soil is much higher than that in the soil submerged by a water layer. The influence of the water layer on the soil is confined to a depth of 3—4 cm. The high content of oxygen beneath the cultivated layer is due to the fact that the soil is unsaturated with water in these low–lying layers.

The oxygen content of the soil increases when the field is drained at the middle stage of rice growth[1]. Hence it is a common agricultural practice to regulate the water and air regime of the soil by taking various irrigation or drainage measures, so as to control some physico–chemical properties of the soil and rice growth.

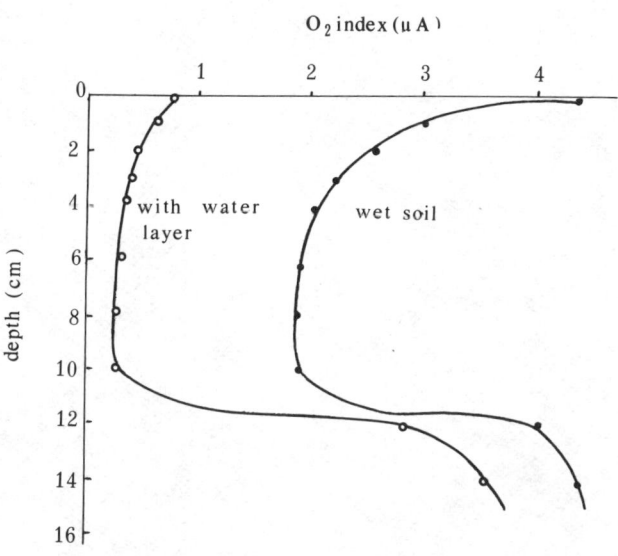

Fig. 3–20 Influence of irrigation mode on O_2 content of paddy field (mountain region of Jiangxi) (measured with micro–Pt electrode)[2]

REFERENCES

(1) Institute of Soil Science, 1961. Soil Environment of High–yield Rice. Chapter 6. Science Press, Beijing.
(2) Institute of Soil Science, 1978. Soils of China. Part B, Chapter 10. Science Press, Beijing.

(3) Yu Tian-ren, Zhang Xiao-nian et al., 1980. Electrochemical Methods and Their Applications in Soil Research. Chapter 15. Science Press, Beijing.
(4) Stolzy, L.H. and Letey, L., 1964. Measurement of oxygen diffusion rates with the platinum microelectrode. III. Correlation of plant response to soil oxygen diffusion rates. Hilgardia, 36 (20): 567;576.

CHAPTER 4

IRON AND MANGANESE

BAO XIE-MING

One of the most important changes in paddy soils induced by the change of oxidation–reduction regime during alternate wetting and drying is the change in the forms of iron and manganese. The oxidation or reduction of iron and manganese deserves special attention, because it is easily noticcable morphologically and plays an important role in soil genesis and plant growth, although it is caused passively by other factors, such as oxygen or organic reducing substances.

The behavior of iron and manganese is similar in many ways with respect to oxidation–reduction. However, there are also differences. Since in paddy soils iron is higher in amount and more important in significance than manganese, in this chapter special attention is paid to the discussion of iron.

4.1 CHEMICAL BEHAVIOR OF IRON AND MANGANESE IN SOIL

The oxidation or reduction of iron and manganese in paddy soils is affected by pe, pH and O_2 or CO_2 etc. of the soil. These will be discussed respectively.

4.1.1 Effect of pe and pH

It is known from Table 1–1 of Chapter 1 that in oxidation–reduction reactions of iron and manganese there is not only the transfer of electrons but also the participation of proton. Taking the reduction of Fe_2O_3 as an example:

$$Fe_2O_3 + 6H^+ + 2e^- \rightleftharpoons 2Fe^{2+} + 3H_2O \tag{4-1}$$

If the equilibrium constant log K is 25.0, it follows:

$$pFe^{2+} = pe + 3pH - 12.5 \tag{4-2}$$

It can be seen from the above equation that under definite conditions the concentration of Fe^{2+} in reduced soils is quantitatively related to the pe and pH of the medium.

The transformation of manganese in soils is also affected by the pe and pH. For the reduction of MnO_2:

$$MnO_2 + 4H^+ + 2e^- \rightleftharpoons Mn^{2+} + 2H_2O \tag{4-3}$$

the relation is:

$$pMn^{2+} = 2pe + 4pH - 41.7 \tag{4-4}$$

4.1.2 Effect of O_2 or CO_2 partial pressure

From the viewpoint of chemical equilibrium O_2 functioning as an electron acceptor can directly affect the direction of transformation and the equilibrium position of iron and manganese. For the reduction of O_2:

$$O_2 + 4H^+ + 4e \rightleftharpoons 2H_2O \tag{4-5}$$

combining Equations (4–5) and (4–1), we get:

$$Fe_2O_3 + 4H^+ \rightleftharpoons 2Fe^{2+} + 2H_2O + 1/2\ O_2(gas) \tag{4-6}$$

$$K = \frac{(Fe^{2+})^2 (P_{O_2})^{1/2}}{(H^+)^4} = 10^{-16.5} \tag{4-7}$$

Taking logarithmic form and after rearrangement:

$$pFe^{2+} = 1/4 \log P_{O_2} + 2pH + 8.25 \tag{4-8}$$

For the reduction of MnO_2 we can get a similar equation.

Thus it can be seen that the higher the partial pressure of O_2 in the air, the lower will be the concentration of Fe^{2+} or Mn^{2+}.

In submerged soils the concentration of carbon dioxide may be several hundred times higher than that in the atmosphere. Since the solubility product of ferrous carbonate and manganous carbonate is very small, the effect of CO_2 on the precipitation–solution equilibrium of iron and manganese should not be overlooked.

Considering the solution of $FeCO_3$:

$$FeCO_3 \rightleftharpoons Fe^{2+} + CO_3^{2-} \tag{4-9}$$

The solubility product of $FeCO_3$ is 3.5×10^{-11}. If the partial pressure of carbon dioxide is 0.1 atmosphere, the concentration of CO_3^{2-} at pH 7 will be about 6×10^{-5} mole per liter. This concentration is high enough to induce the precipitation of Fe^{2+} at a concentration of as low as 6×10^{-7} mole per liter. However, following the further increase in CO_2 concentration some $FeCO_3$ can be dissolved as bicarbonate.

The relation between Mn^{2+} concentration and CO_2 partial pressure is

$$pMn^{2+} = \log P_{CO_2} + 2pH - 8.2 \tag{4-10}$$

4.1.3 Sensitivity to oxidation–reduction condition and acidity

From the above paragraphs it is clear that there are similarities in the sensitivity of iron and manganese to the oxidation–reduction condition. On the other hand, since the standard oxidation–reduction potential E^0 in the reaction

$$MnO_2 + 4H^+ + 2e \rightleftharpoons Mn^{2+} + 2H_2O \tag{4-3}$$

is higher than the E^0 in the reaction

$$Fe(OH)_3 + 3H^+ + e \rightleftharpoons Fe^{2+} + 3H_2O \tag{4-11}$$

the two elements will respond to the oxidation–reduction condition differently. According to the sequential reduction mentioned in Chapter 1, manganese compounds in soils should be more easily reducible than iron compounds. Actually this is just the case. In Table 4–1 are shown the amounts of iron and manganese reduced by hydroquinone in different horizons of a paddy soil derived from red soil. It can be seen that although the amount of reduced iron increases slightly with the increase in concentration of the reducing agent, it accounts for the total iron by not more than several thousandths. On the contrary, with the exception of the glei horizon the amount of manganese reducible by hydroquinone accounts for a large proportion (with a maximum of 70%) of the total

4.1 Chemical behavior of iron and manganese in soil

manganese of the soil. The amount of extractable manganese attains a high value at a very low concentration of hydroquinone. This means that manganese compounds possess a strong oxidizing property as compared with iron compounds. It is for this reason that at ordinary soil pH ferrous iron can exist only at a low Eh, whereas manganous manganese can exist even in well-aerated soils.

TABLE 4-1 Easily reducible iron and manganese contents in different horizons of a paddy soil (derived from red soil)[3]

Horizon	Concentration of hydroquinone (%)	Easily reducible (mg/100g)		In total (%)	
		Fe	Mn	Fe	Mn
A	0.02	8.0	10.6	0.27	23.7
	0.05	9.3	18.0	0.32	40.1
	0.1	11.0	19.0	0.37	42.4
	0.2	13.5	18.3	0.46	40.9
	0.4	14.5	18.5	0.49	41.3
B	0.02	1.3	60.4	0.04	68.0
	0.05	2.0	62.8	0.07	70.8
	0.1	1.9	62.8	0.07	70.8
	0.2	2.9	60.4	0.10	68.0
	0.4	4.8	62.7	0.16	70.7
G	0.02	1.7	tr.	0.10	
	0.05	1.9	tr.	0.11	
	0.1	2.9	tr.	0.16	
	0.2	2.7	tr.	0.16	
	0.4	2.4	tr.	0.13	
C	0.02	2.0	16.4	0.06	42.3
	0.05	2.1	16.5	0.06	42.4
	0.1	2.8	14.0	0.07	36.1
	0.2	3.5	18.0	0.09	46.4
	0.4	3.3	16.5	0.09	42.6

The pK_{sp} of Mn(OH)$_2$ in water is about 13, much smaller than that of Fe(OH)$_3$ or Fe(OH)$_2$. It is for this reason that iron is more easily precipitated than manganese. However, since the content of iron in soils is much higher than that of manganese, the amount of dissolved iron may be quite high, provided that sufficient hydrogen ions are present in the system. If we take the ease of solution by sulfuric acid as a relative index of the activity of iron or manganese, it can be seen from comparisons between Tables 4-2 and 4-1 that the amount of acid-extractable manganese is roughly comparable to that of reducible manganese, and increases only to a small extent with the increase in acid concentration, whereas the amount of extractable iron increases greatly with the increase in acid concentration, and can attain a value of several hundred milligrams (corresponds to 20% of the total iron) if the acid concentration is sufficiently high. These results demonstrate clearly that the sensitivity of iron to the solution action of acid is more remarkable than that of manganese.

TABLE 4-2 Acid-soluble iron and manganese contents in different horizons of a paddy soil (derived from red soil)[3]

Horizon	Concentration of H_2SO_4 (N)	Acid-soluble (mg/100g)		In total (%)	
		Fe	Mn	Fe	Mn
A	0.05	240	16.0	8.0	36.0
	0.1	264	14.0	8.8	31.2
	0.5	480	15.0	16.0	33.5
	1	480	14.0	16.0	31.2
	5	648	14.0	21.6	31.2
B	0.05	80	8.6	2.7	7.7
	0.1	80	15.0	2.7	16.9
	0.5	292	24.0	10.0	27.1
	1	472	35.0	16.2	39.5
	5	632	56.6	21.6	63.0
G	0.05	52	tr.	3.0	
	0.1	64	tr.	3.6	
	0.5	100	tr.	5.7	
	1	124	tr.	7.0	
	5	260	tr.	14.6	
C	0.05	64	5.0	1.6	12.9
	0.1	264	10.0	6.7	25.8
	0.5	488	12.5	12.3	32.5
	1	562	16.0	14.2	41.3
	5	616	19.0	15.6	49.0

4.2 IRON IN PADDY SOILS

4.2.1 Amount of active iron

The term "active iron" is generally used to denote that part of iron in the soil which can be dissolved by reducing agents, complexing agents or acids. Since the extracting power of various methods is different, the extracted amount varies to a large extent. It is seen from the results of Table 4–3 that the amount of active iron extracted by alternate treatments with 4% $Na_2S_2O_4$ and 0.05 N HCl accounts for 50—70% of the total iron, that extracted by 0.5 N H_2SO_4 accounts for 1—9% of the total, and that extracted by $(NH_4)_2C_2O_4$ is close to that extracted by H_2SO_4. The amount of ferrous iron under natural submerged conditions accounts for several hundredths of the total iron, and is of the same order of magnitude as that extracted by 0.5 N H_2SO_4 or $(NH_4)_2C_2O_4$.

TABLE 4–3 Amount of active iron in paddy soils as extracted by different methods[2, 3, 8]

Soil	Depth (cm)	Total Fe_2O_3 (%)	Active iron		Extractant
			Fe_2O_3(%)	In total (%)	
Gleyed paddy soil (Wuxing)	0—14	4.55	0.45	9.9	Oxalic acid-oxalate
	14—24	4.27	0.46	10.8	
	24—40	4.05	0.11	2.7	
	40—70	2.48	0.07	2.8	
	70—100	7.35	0.32	4.4	
Paddy soil from red soil (Dongxiang)	0—15	3.41	0.31	9.1	0.5N H_2SO_4
	15—52	5.31	0.12	2.3	
	52—58	3.15	0.04	1.3	
	58—	7.13	0.16	2.2	
Paddy soil from red soil (Jinxian)	0—11	3.52	2.03	57.7	Intermittent 4% $Na_2S_2O_4$ and 0.05N HCl
	11—17	6.35	4.66	73.4	
	17—29	5.59	3.76	67.3	
	29—60	6.45	3.42	53.0	

The activity of iron in paddy soils is related to the form of iron compounds. Usually the stronger the dehydration or the more the crystallization, namely the more "aged" the iron oxides, the lower the activity will be. From experimental results of the effect of alternate wetting and drying on the activity of iron shown in Table 4-4 it can be seen that the amount of reducible or acid-soluble iron decreases remarkably after treatment, especially those treated at high temperature (120°C). Apparently, this is due to the effect of dehydration.

TABLE 4-4 Effect of alternate drying and wetting on the activity of iron of the soil[3]

Soil	Extractant	Iron (mg Fe/100g)		
		CK	Air-drying	Oven-drying
Gleyed paddy soil	$NH_4Ac+0.1\%$ hydroquinone	17.2	17.0	14.6
	$0.05N\ H_2SO_4$	263	130	103
	$0.5N\ H_2SO_4$	592	438	320
Paddy soil from red soil	$0.05N\ H_2SO_4$	52.0	14.4	8.8
	$0.5N\ H_2SO_4$	126	72.8	40.8
Red soil	$NH_4Ac+0.1\%$ hydroquinone	2.0	2.4	1.0
	$0.05N\ H_2SO_4$	56.8	32.0	15.2
	$0.5N\ H_2SO_4$	184	130	68.0

4.2.2 Forms of ferrous iron

In submerged paddy soils a part of the iron compounds is reduced to ferrous iron. Under strongly reducing conditions the amount of ferrous iron may be as high as 4—5 hundred milligrams per 100g of soil. This ferrous iron is the most active part of the iron compounds. This part of iron may be distinguished into the water-soluble, the exchangeable, the complexed (with the solid phase of the soil) and the precipitated. In Table 4-5 is shown the general pattern of the distribution of the four forms of ferrous iron. In the following, the amounts of various forms of ferrous iron and factors affecting them, and the chemical equilibria among them will be discussed.

4.2.2.1 *Water-soluble*

Water-soluble ferrous iron is the most active part of ferrous iron. It is closely related to the ionic composition of soil solution, the migration of nutrients and the growth of plants. It may be further distinguished into the ionic and the chelated.

The amount of water-soluble ferrous iron in soils is determined chiefly by the organic matter content and pH of the soil. Organic matter can either affect

4.2 Iron in paddy soils

the oxidation–reduction condition of the soil and thus the total amount of ferrous iron, or function as a chelating agent for water–soluble ferrous iron. From the data shwon in Table 4–5 it is seen that for the untreated soil the amount of water–soluble iron accounts for less than 1% of the total ferrous iron, whereas for the soil with added organic matter it may be as high as 20—25% of the total.

TABLE 4–5 Distribution of various forms of ferrous iron in paddy soils under submerged condition[7]

Soil	O.M. added (%)	pH	Eh (mV)	Ferrous iron (mg/100g)	Distribution of ferrous iron (%)			
					Soluble	Exchange-able	Complexed	Precipitated
Paddy soil from yellow–brown soil	0	6.29	274	55.9	0.04	0.4	36.0	63.7
	1	6.20	263	98.0	0.42	5.3	31.7	62.6
	3	5.98	241	112	0.98	14.0	27.6	57.4
Yellow–brown soil	0	7.50	260	37.0	—	1.5	21.6	76.8
	1	6.10	225	94.5	0.2	1.8	23.6	74.4
	3	5.65	216	112	8.8	6.8	22.3	61.7
	5	5.55	192	169	13.2	21.3	18.9	46.6
Paddy soil from red soil	0	5.54	381	15.2	1.0	0.5	23.2	75.0
	1	5.24	170	43.3	25.6	6.9	—	—
	3	4.95	−116	125	11.8	5.0	—	—
Gleyed	0	7.30	48	443	0.1	1.4	15.5	83.1

The amount of water–soluble ferrous iron is strongly affected by the pH of the medium. This is because the solubilities of various forms of precipitated ferrous iron such as $Fe(OH)_2$ and FeS are closely related to pH. In Fig. 4–1 is shown an experimental result in this respect. In the figure the theoretical solubility curves are also given for comparison, assuming the solubility product of $Fe(OH)_2$ as 6.4×10^{-18}, and that of FeS as 3.7×10^{-19}. In a simulating experiment the organic matter and free iron oxides of a paddy soil derived from red soil were removed, ferrous iron and sulfide were added in various ratios, and then the pH of the soil was adjusted.

It is demonstrated in the figure that the relationship between the concentration of water–soluble ferrous iron of the soil and the pH can be examined in two pH ranges. At pH of higher than 5—5.5 the change in pFe^{2+} caused by one unit change of the pH is about 0.7—0.9 for the three soils, and in the simulating experiment it may be expressed by an empirical equation:

$$pFe^{2+}=1.47pH-4.88 \qquad (4-12)$$

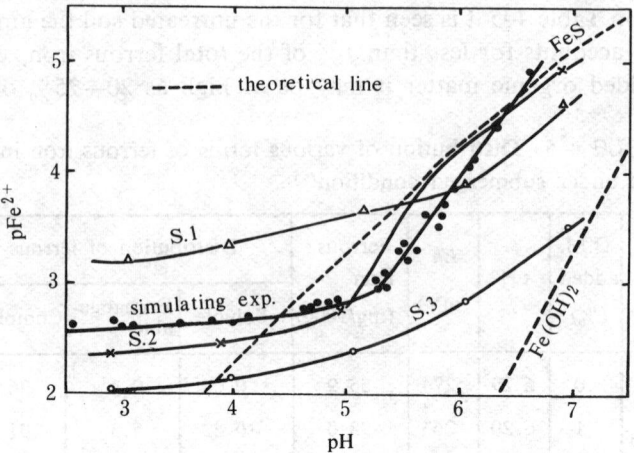

Fig. 4-1 Relationship between water-soluble ferrous iron and pH of paddy soils[7] (S. 1, derived from red soil; S. 2, derived from yellow soil; S. 3, gleyed)

At lower pH, the slope is smaller. The actual concentration of ferrous iron in soil solution lies within the range between those for FeS and $Fe(OH)_2$. It should also be noticed that at a same pH the concentration of water-soluble ferrous iron in various soils differs greatly. The concentration in a gleyed paddy soil is especially high, due presumably to the presence of water-soluble ferrous chelates.

TABLE 4-6 Amount of water-soluble ferrous iron in paddy soils (Fujian Province)[10]

Soil type	pH	O.M.(%)	Ferrous iron (ppm) (determined in laboratory)			Ferrous iron (ppm) (determined in field)
			Ionic	Chelated	Total	
Ordinary	5.8	3.30	18.7	6.9	25.6	27.6
Ordinary	5.9	—	12.6	3.9	16.5	19.5
Ordinary	6.2	—	18.1	1.7	19.8	13.5
Gleyed	6.4	3.84	15.6	7.4	23.0	48.6
Gleyed	6.3	4.70	14.6	2.0	16.6	27.6
Gleyed	5.0	5.37	29.3	4.6	33.9	75.8
With rusty water	6.3	3.04	97.5	13.5	111	105
With rusty water	6.5	4.95	26.7	18.1	44.8	44
With rusty water	6.2	19.7	105	81.7	187	126
Acid sulfate	2.9	—	627	32	659	358
Acid sulfate (reclaimed)	6.4	—	69.4	5.1	74.5	74.9

4.2 Iron in paddy soils

In Table 4-6 is shown the amount of water-soluble ferrous iron under field conditions for some representative paddy soils. According to statistics from some analytical data, the average amount of water-soluble ferrous iron in ordinary paddy soils is 23 ppm; in gleyed paddy soils, 46 ppm; in paddy soils with rusty water, 108 ppm; and in acid sulfate soils, up to several hundreds ppm. In water-soluble ferrous iron the ionic form accounts for more than 70%, with the chelated form 10—30%. However, in some soils rich in organic matter the chelated iron may account for 40% of the total water-soluble iron.

The chelated water-soluble ferrous iron may be further characterized by the ion-exchange resin method as those carrying negative charge and those carrying positive charge. Their relative proportions in different soils vary greatly (see Table 4-7).

TABLE 4-7 Fractionation of chelated ferrous iron according to electric charge[9]

Soil	Locality	Treatment	Ferrous iron (ppm)		
			Positive charge	Negative charge	Sum
With rusty water	Shunchang	Fresh soil	72.7	9.7	82.4
With rusty water	Shunchang	Fresh soil	5.5	6.0	11.5
Gleyed	Shunchang	Fresh soil	2.3	2.2	4.5
Paddy from red soil	Jinxian	5% vetch	8.3	5.4	13.7
Gleyed	Xinghua	5% vetch	15.3	0.4	15.7
Sandy	Zixi	5% vetch	1.1	0.2	1.3

It has been measured by means of the potentiometric titration method that the stability constant (mean apparent stability constant) $\log K$ of chelated ferrous iron ranges between 2.4—4.0[10], depending on the kind of organic matter. For the same green manure, the stability of Fe^{2+}-chelates is highest at the period of intensive decomposition of organic matter.

The ratio of ionic iron to chelated iron is determined by the chelation-dissociation equilibrium of iron-chelates. This reaction obeys the mass-action law. It can be seen from Fig. 4-2 that under conditions of constant concentration of ferrous iron, the larger the amount of chelating agent (in this case the incubation solution of green manure), the higher the proportion of ionic iron. And, in the presence of sufficient chelating agents nearly all of the ferrous iron is present in the chelated form. On the other hand, if the amounts of ferrous iron and chelating agents are kept constant other ions can compete for chelating groups with ferrous iron, liberating a part of the chelated iron to ionic form. In this case the extent of decrease in proportion of chelated iron is affected by the kind and quantity of these ions as well as by the stability of chelates formed between these ions and the chelating agents. For example, if Mn^{2+} ions are added to the soil solution or the incubation solution of green manure the amount of chelated ferrous iron will decrease (Fig. 4–3). However, due to some unknown reasons, one milligram of Mn^{2+} can liberate only 0.4—0.8 milligrams of Fe^{2+}.

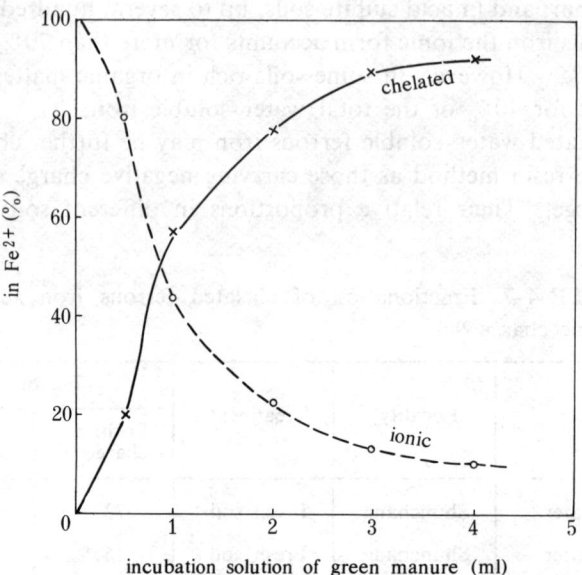

Fig. 4–2 Relationship between percentage of chelated ferrous iron and amount of chelating agents[10]

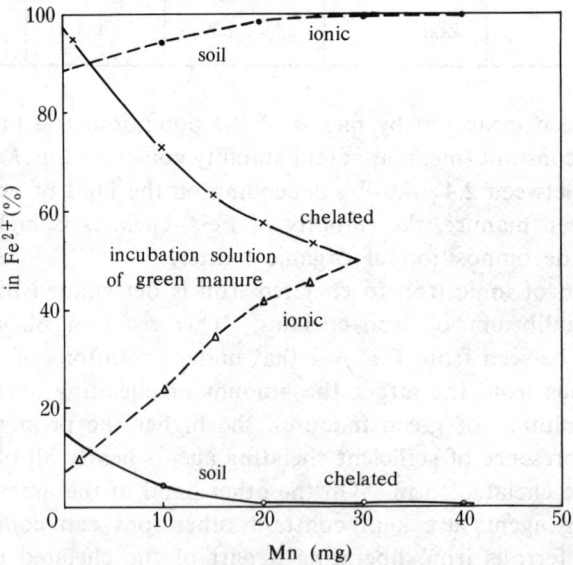

Fig. 4–3 Effect of manganous ions on chemical equilibrium of chelated ferrous iron[9, 10]

4.2 Iron in paddy soils

Hydrogen ions have a strong influence on the chelation–dissociation of ferrous iron. It is evident from Fig. 4–4 that for the soil solution of three soils there is a linear relationship between the pH and the logarithm of the amount of ionic or chelated ferrous iron. The increase in amount of ionic iron with the fall of the pH is due apparently to the competition of hydrogen ions for chelating groups with ferrous ions.

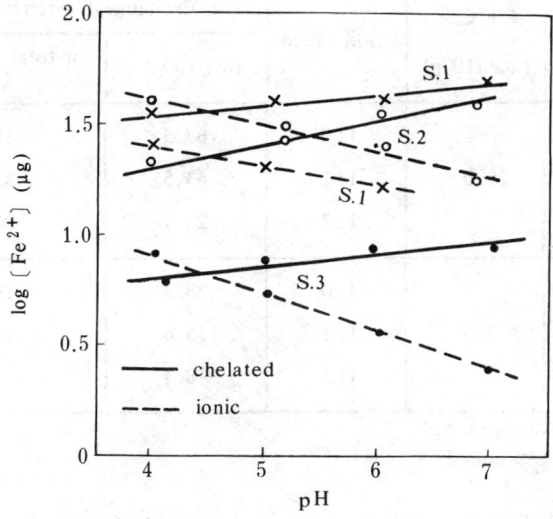

Fig. 4–4 Effect of pH on chemical equilibrium of chelated ferrous iron[10]
(S. 1, sandy soil; S. 2, derived from red soil; S. 3, gleyed)

4.2.2.2 Exchangeable

The amount of exchangeable ferrous iron in paddy soils increases with the intensification of the reduction condition and the fall of the pH. Besides, the cation–exchange capacity of the soil is also an important factor. For, under otherwise constant conditions, the larger the cation–exchange capacity of the soil the larger the possibility of the existence of exchangeable ferrous iron. It can be seen from an experiment shown in Table 4–8 that following the addition of quartz sand and consequently the decrease in the cation–exchange capacity of the soil, the amount of exchangeable ferrous iron decreased regularly. It is also noticeable that the amount of exchangeable ferrous iron and its proportion in the total ferrous iron for a paddy soil derived from yellow–brown soil were much higher than those for a paddy soil derived from red soil. This is related to the higher cation–exchange capacity for the former soil.

Sulfide ions can compete for ferrous iron with the cation–exchange sites of the soil to form FeS precipitate, thus causing the quantity of exchangeable ferrous iron to decrease. Results from an experiment showed that for a submerged soil in which the sulfur had been removed previously the amount of exchangeable ferrous iron at pH 7 was 1.8—2.6 m.e./100g, corresponding to 21—26% of the total ferrous iron, but it decreased regularly with the increase in amount of added S^{2-} (Table 4-9).

TABLE 4–8 Relationship between exchangeable ferrous iron and the cation–exchange capacity of paddy soils[7]

Soil	C.E.C. (m.e./100g)	Soil : sand	Exchangeable ferrous iron	
			(mg/100g)	In total ferrous iron (%)
From yellow-brown soil*	21.0	1 : 0	67.5	38.6
		1 : 1	63.5	32.4
		1 : 3	23.9	20.1
From red soil**	7.36	1 : 0	23.3	19.4
		1 : 1	15.6	11.6
		1 : 3	8.4	5.0

* 200 mg/100g of Fe^{2+} added
** 150 mg/100g of Fe^{2+} added

TABLE 4–9 Effect of sulfide ions on exchangeable ferrous iron in reduced soils (pH 7.0 ± 0.1)[7]

Soil	Total Fe^{2+} (mg/100g)	Na_2S added (m mole/100g)	Exchangeable Fe^{2+}	
			(mg/100g)	In total Fe^{2+} (%)
Submerged red soil	195	0	50.9	26.1
		10	18.7	9.6
		20	5.8	3.0
		30	1.9	1.0
Gleyed paddy soil	348	0	72.9	21.0
		10	50.3	14.5
		20	20.4	5.9
		30	12.1	2.5

In submerged paddy soils the amount of exchangeable ferrous iron accounts for less than 20% and mostly less than 5% of the total ferrous iron, corresponding to less than 10% of the exchangeable cations. This low percentage of exchangeable

4.2 Iron in paddy soils

iron in submerged soils is due to the competition of S^{2-} ions and OH^- ions for ferrous iron with the cation–exchange sites, although ferrous iron itself has a high bonding energy with soil particles. Under natural conditions of submerged paddy soils the pH is generally higher than 6, and there is always the presence of a certain amount of sulfide ions. These are the basic reasons why the content of exchangeable ferrous iron in submerged paddy soils is not high.

4.2.2.3 Complexed and precipitated

The term "complexed" ferrous iron is used here to distinguish it from that part of ferrous iron which is chelated with the water–soluble organic substances in that it is associated with the organic matter of the solid phase. Actually, however, not all of the water–soluble "chelated" ferrous iron is bound by the mechanism of chelation, and some of the organic matter in the solid phase may also contain chelating groups. The distinction used here is primarily in phase rather than in nature.

In paddy soils the complexed ferrous iron usually accounts for 15—35% of the total ferrous iron. The amount of complexed iron is closely related to the content of organic matter of the soil. It was found that complexed ferrous iron in soils increased in amount after the addition of organic matter, and disappeared when all of the soil organic matter were removed. For the same type of soil the amount is related to the organic matter content almost linearly (Fig. 4–5). It can be calculated from the slope of the straight line that one gram of soil organic matter can complex about 6—8 mg of ferrous iron.

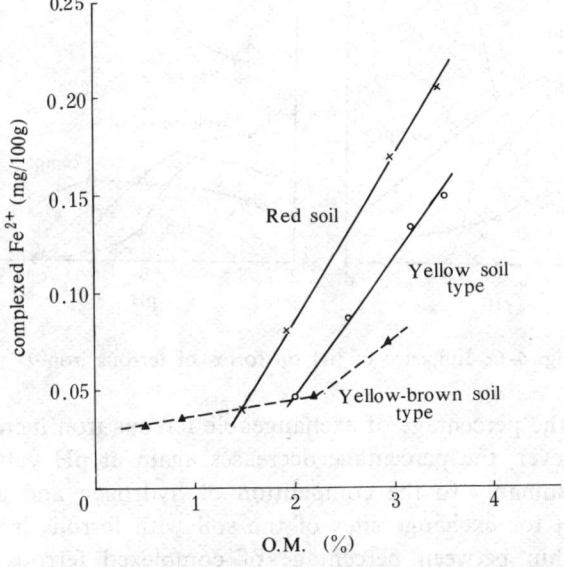

Fig. 4–5 Relationship between complexed ferrous iron and organic matter content of the soil[6]

Precipitated iron occupies the most important portion of ferrous iron, accounting for 60—80% of the total. In paddy soils containing a large amount of sulfide the proportion of precipitated ferrous iron is higher. The proportion decreases after the addition of easily decomposable organic matter due to the pronounced increase in water-soluble and exchangeable iron, although the absolute amount of precipitated iron increases.

4.2.3 Physico-chemical equilibria among various forms of ferrous iron

There is a dynamic equilibrium among various forms of ferrous iron. This equilibrium is affected by the total amount of ferrous iron, chemical composition of the soil and characteristics of the solid phase. Besides, environmental conditions may also have some bearing on it. Reasoning from theoretical considerations, the pH of the medium should have a significant influence on the equilibrium. This is illustrated in Fig. 4-6. The increase in percentage of precipitated ferrous iron and the decrease in percentage of water-soluble iron with the increase in pH are due to the fact that the solubilities of all of Fe^{2+} precipitates such as FeS, $Fe(OH)_2$ and $FeCO_3$ are pH-dependent.

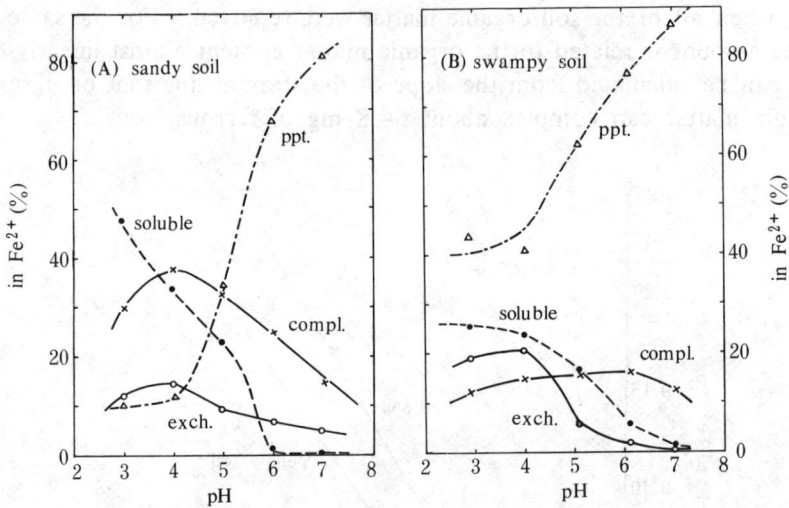

Fig. 4-6 Influence of pH on forms of ferrous iron[7]

In the figure the percentage of exchangeable ferrous iron increases with the fall of pH. However, the percentage decreases again at pH values lower than about 4, due presumably to the competition of hydrogen and aluminum ions present at low pH for exchange sites of the soil with ferrous iron.

The relationship between percentage of complexed ferrous iron and pH is similar to that between exchangeable iron and pH. At a low pH hydrogen and aluminum ions can also compete for active groups of complexing agents, decreasing the percentage of complexed ferrous iron.

4.3 Manganese in paddy soils

It can also be seen from Fig. 4–6 that the absolute values of the percentage of various forms of ferrous iron in two soils are different. For instance, in the swampy paddy soil the percentage of precipitated ferrous iron is higher than that in the sandy paddy soil at the same pH. It has been known that the former soil contains a large amount of sulfides. The percentages of exchangeable ferrous iron are also different for the two soils because of the difference in cation–exchange capacity.

The speed of chemical reactions among various forms of ferrous iron is quite high. In an experiment it was found that after the removal of water–soluble, exchangeable and complexed ferrous iron from the soil the transformation from precipitated iron to water–soluble or exchangeable iron attained a steady state within one hour and to complexed iron within four hours, if the soil was kept under room temperature and isolated from atmospheric air (Table 4–10). It is conceivable that the transformation of other forms of ferrous iron to precipitated iron would need a still shorter time.

TABLE 4–10 Transformation rate of precipitated ferrous iron[7]

Soil	Reaction time (hr.)	Ferrous iron (mg/100g)				
		Soluble	Exchangeable	Complexed	Precipitated	Sum
Submerged red soil	1	0.40	0.81	7.50	7.90	16.61
	4	0.32	0.81	9.87	5.32	16.32
	8	0.32	0.51	10.1	5.00	15.93
Loamy paddy soil	1	0.08	0.41	1.72	1.80	4.01
	4	0.08	0.41	2.46	1.31	4.26
	8	0.08	0.41	2.48	1.30	4.27

4.3 MANGANESE IN PADDY SOILS

Manganese in soils can be distinguished as the water–soluble, exchangeable, easily reducible and difficultly soluble. The manganese, easily reduced to a divalent state by reducing agents such as hydroquinone, is chiefly in the form of manganese oxides. Manganese which is difficult to dissolve (difficultly soluble) is in the form of inactive minerals.

In well–drained paddy soils manganese is present mainly in the form of manganic oxides. Like ferric oxides, these manganic oxides have a very low solubility in water. However, these oxides are more easily reducible than ferric oxides. Since manganous ion resists air oxidation more than ferrous iron, there can be a certain amount of the ion existing in well–drained soils.

4.3.1 Amount of active manganese

The term "active" manganese is generally used to denote that part of manganese compounds which can be extracted by reducing agents (or sometimes including that extracted by a complexing agent). This part of manganese undergoes transformation in the course of alternate wetting and drying of the paddy field, hence is the more important part of manganese in paddy soils. Its amount in different soils varies widely. Taking a paddy soil derived from red soil as an example, it can be seen from Table 4–1 that in the illuvial horizon the amount attains a value of 60mg/100g, accounting for 70% of the total manganese, whereas in the gleyed horizon the amount is very low. In paddy soils of North China the amount is usually several ten milligrams per 100 grams of soil, accounting for 30—60% of the total manganese. In neutral paddy soils the amount is usually of the same order of magnitude as in North China.

Like iron compounds, the activity of manganese compounds in soils is related to the degree of ageing (dehydration). From an experiment shown in Table 4–11 it can be seen that the activity of manganese compounds decreased after the treatment of alternate wetting and drying.

TABLE 4–11 Effect of alternate drying and wetting on the activity of manganese of the soil[3]

Soil	Extractant	Manganese (mg Mn/100g)		
		CK	Air–drying	Oven–drying
Gleyed paddy soil	$NH_4Ac + 0.1\%$ hydroquinone	5.1	4.0	3.1
	$0.05N\ H_2SO_4$	3.8	4.4	3.1
	$0.5N\ H_2SO_4$	4.2	4.7	3.9
Paddy soil from red soil	$NH_4Ac + 0.1\%$ hydroquinone	35.4	24.0	10.6
	$0.05N\ H_2SO_4$	28.8	20.7	16.0
	$0.5N\ H_2SO_4$	34.6	36.6	23.6
Red soil	$NH_4Ac + 0.1\%$ hydroquinone	12.0	12.0	6.4
	$0.05N\ H_2SO_4$	14.0	6.2	5.8
	$0.5N\ H_2SO_4$	14.8	6.6	7.4

4.3.2 Relationship between amount of exchangeable manganese and pH

It has been mentioned in the above that in paddy soils there may be a certain

4.3 Manganese in paddy soils

amount of exchangeable (including water-soluble) manganese even under well-aerated conditions. In the cultivated layer of acid paddy soils the amount is usually 2—9 mg per 100 grams of soil, with a maximum of 20 mg, corresponding to 0.7 m.e./100g. Therefore, exchangeable manganese is of certain significance in acid soils.

Soil pH has a remarkable effect on the amount of exchangeable manganese. The general tendency is that the higher the pH, the lower the content of exchangeable manganese. Ordinarily, in soils with a pH of higher than 6.5—7.0 the content of exchangeable manganese is very low. According to some experimental results, the negative logarithm of the amount of exchangeable manganese when expressed as m M/100g soil is linearly related to the pH of the soil. It can be seen from examples shown in Fig. 4–7 that the relationship can be expressed by equation (4–13):

$$pMn^{2+} = a\ pH - b \tag{4-13}$$

This equation is similar in form to Equation (4–4) or (4–10).

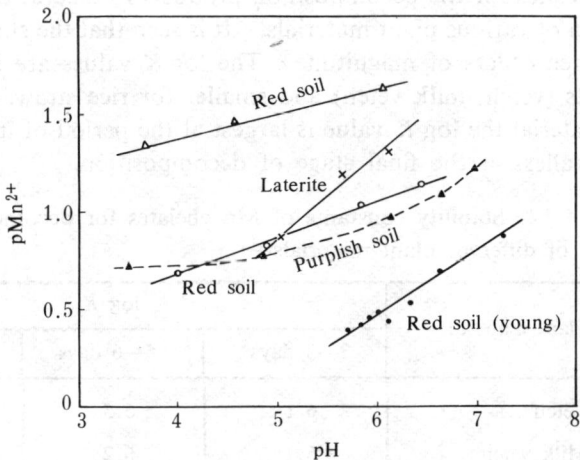

Fig. 4–7 Relationship between exchangeable manganese and pH of paddy soils[4, 5] (pMn^{2+} denote negative logarithm of Mn^{2+} in m mole per 100g of soil)

When $pMn^{2+} = 0$, the numeral value of pH is b/a. This means that b/a is the pH at which one hundred grams of soil contain one millimole of exchangeable manganese. Of course, this value is difficult to attain due to the insufficiency in readily reducible manganese compounds in the soil and other factors, and hence the b value has little practical significance. However, the b value can be used as a relative index for evaluating the level of active manganese in different soils. It can be seen from Fig. 4–7 that the pMn^{2+} values at the same pH for different soils vary greatly. The pMn^{2+} for a young paddy soil is particularly small. Actually, this soil is a Quaternary red clay being cultivated just for rice, and contains a high amount of active manganese.

A higher "a" value means a greater change in the amount of exchangeable manganese caused by a one unit change of soil pH. The cation–exchange capacity of the soil should be an important factor in affecting the "a" value. For some soils (such as a paddy soil derived from sandstone) the amount of exchangeable manganese increases very little when the pH decreases beyond a certain limit. This is due to the insufficiency of active manganese of the soil. Therefore, the quantitative relationship between pH and amount of exchangeable manganese is conditional.

The effect of pH on the amount of exchangeable manganese is of practical significance. This will be discussed in Chapter 10.

4.3.3 Chelation with organic substances

Like ferrous iron, manganous ions can also be chelated by some organic substances of the soil. The stability constant of Mn–chelates is related to the nature of the chelating agents. In Table 4–12 are shown the stability constants $\log K$ of Mn–chelates for the decomposition products formed at different periods of decomposition of various plant materials. It is seen that the stability constants may vary by three orders of magnitude. The $\log K$ values are larger for leguminaceous plants (vetch, milk vetch) and smaller for rice straw. For the same kind of plant material the $\log K$ value is largest at the period of intensive decomposition and smallest at the final stage of decomposition.

TABLE 4–12 Stability constants of Mn–chelates for decomposition products of different plant materials[11]

Treatment	Plant material	$\log K$		
		2—3 days*	5—6 days	Over 20 days
CK	Vetch	6.1	6.5	3.8
	Milk vetch	6.1	6.2	3.4
	Rice straw	4.7	4.8	3.9
Kaolinite	Vetch	4.9	5.0	3.5
	Milk vetch	4.4	4.9	3.5
	Rice straw	4.1	4.3	
Red soil	Vetch	3.6	3.3	
	Milk vetch	3.1	3.3	
	Rice straw	2.0	3.0	

* Incubation time

It can also be seen from the table that if kaolinite was added to the suspension before incubation the stability constants were smaller. This implies that

4.3 Manganese in paddy soils

there is the possibility of adsorption of chelating agents by clay. This supposition was supported by an experiment in which the decrease in stability constants was greater if red soil instead of kaolinite was added to the suspension: the more the amount of soil added the greater the decrease of the constant. The adsorption of chelating agents by soil particles seems to be related to the iron oxides carrying positive charge, for it was found that the higher the content of active iron oxides in the added soil, the smaller the log K value (Table 4–13).

TABLE 4–13 Decrease in stability constant in relation to iron content of the soil

Locality	Active Fe_2O_3 (%)	log K_R*	log K_M**
Jiangxi	0.90	3.9	3.7
Zhejiang	1.53	2.3	2.7
Nanjing	1.89	2.5	2.8
Jiangxi	4.28	2.3	2.3
Guangdong	10.65	2.2	2.2

* Concentration of chelating agents determined by titration with $KMnO_4$
** Concentration of chelating agents determined by maximum chelating capacity method

When the chelating agents were fractionated by ion–exchange resins, it was found that the chelating power of the fraction carrying negative charge was about the same as that of the unfractionated solution (Table 4–14), whereas, with the exception of the incubation solution of alfalfa, the log K values for the fraction carrying positive charge were smaller than that of the unfractionated solution. For the incubation solution of rice straw no chelating agent carrying positive charge was isolated, and the chelating power of the fraction carrying

TABLE 4–14 Stability constants for chelating agents carrying different electric charges

Plant material	log K		
	Unfractionated	Negative charge	Positive charge
Alfalfa	4.4	2.8	3.1
Hair vetch	4.3	4.3	3.0
Crotalaria	4.3	4.0	3.8
Vetch	3.8	3.9	2.9
Grass	3.9	3.7	3.0
Milk vetch	3.9	3.8	2.9
Rice straw	2.8	2.8	Undetectable

negative charge was similar to the original solution. This was probably due to the dominant role of the carboxyl group in the chelating reaction in this case, because as shown by the infrared spectrogram, for the incubation solution of rice straw there was only a strong —COOH absorption peak.

The coordination number of the ligands in incubation solutions of plant materials was generally around 1.

4.4 DYNAMICS OF IRON AND MANGANESE IN PADDY SOILS

4.4.1 Total amounts of reduced iron and manganese

In the cultivated layer of paddy soils the total amounts of ferrous iron and manganous manganese increase gradually after submerging due to the reduction of oxides of these elements by organic reducing substances. Then, after a certain period the amounts remain practically unchanged. At the latter stage of rice growth the amounts decrease again. The time required for the attainment of steady amounts is determined by the organic matter status of the soil, soil texture and, especially, soil temperature. If drainage for drying the soil is adapted, the amounts may decrease temporarily. In Fig. 4-8 is shown a representative pattern in this respect. It should also be noticed that the behaviors of iron and manganese showed some differences. For manganous manganese, the time required for the attainment of a steady value was shorter and the range of fluctuation much smaller than for ferrous iron. This can be explained as follows. After submerging the soil, active manganic compounds are reduced first, and ferric compounds are reduced only after most of the former have been consumed in the oxidation of organic reducing substances. Since there is a wide range of

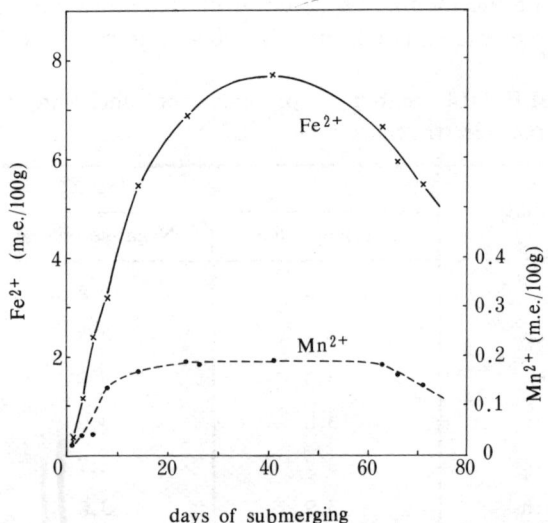

Fig. 4-8 Dynamics of iron and manganese of the cultivated layer of a paddy field during the growing period of rice (Jiangxi, early rice)[1]

4.4 Dynamics of iron and manganese in paddy soils

fluctuation in the amount of reducing substances, and plenty of ferric oxides are available for interactions with organic reducing substances, there is usually a wide range of fluctuation in the amount of ferrous iron in submerged soils.

4.4.2 Water-soluble ferrous iron

Since water-soluble iron is the most active and most important part of ferrous iron, its dynamics in paddy soils deserves special attention. The total amount of water-soluble ferrous iron increases drastically after submerging if a large amount of easily decomposable organic matter is present in the soil. On the other hand, if considered separately, the general patterns of change in amounts of ionic and chelated ferrous iron are quite different. It can be seen from an example shown in Fig. 4–9 that the amount of ionic iron increased rapidly within the first few days of submergence and then came to a steady value (for a sandy paddy soil) or increased continuously (for a paddy soil derived from red soil), whereas the amount of chelated iron went up to the maximum and then decreased again. The chelated iron accounted for 30—90% of the total water-soluble ferrous iron within the first two days of submergence, but decreased afterward. The cause of the change in proportions of the two forms of water-soluble ferrous iron can be imagined as follows: The formation of ferrous iron is the result of reactions between organic reducing substances and ferric oxides. At the beginning of submergence the rate of formation of ferrous iron lags behind that of organic reducing substances, and consequently most of the small amount of ferrous iron appearing at this stage is bound by organic chelating agents. The appearance of a peak in the amount of chelated ferrous iron in the course of submergence is due to the fact that in the period of intensive decomposition of organic matter (about one week after submerging) the amount of chelating agents is at a maximum. In the sandy paddy soil the content of reducible ferric oxides is low, and hence

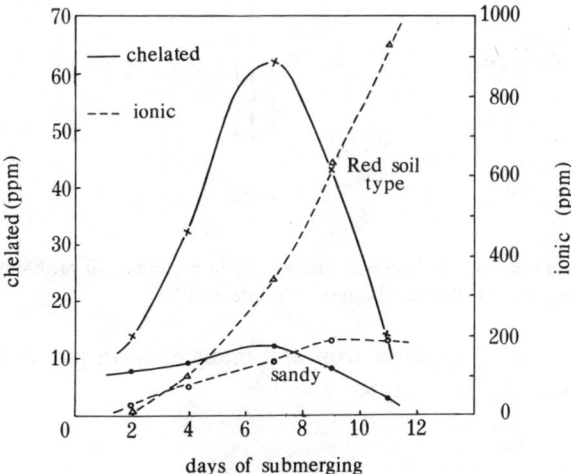

Fig. 4–9 Dynamics of ionic and chelated ferrous iron after submerging of the soil (5% vetch added)[10]

the steady value of ionic ferrous iron is low (190 ppm), whereas in the paddy soil derived from red soil with a high content of ferric oxides the amount of ionic ferrous iron may attain a high value of 900 ppm.

4.4.3 Changes in the profile

In Fig. 4–10 is shown the dynamic change of reduced iron and manganese in a "White soil" profile during the growing period of rice. Except for the periodic change in the cultivated layer, the patterns of change in amounts of divalent iron and manganese in low–lying horizons are quite different. For ferrous iron the amount in the lower part of plowpan (20—23 cm) remained at a low level during the whole submerging period, although in the upper part of plowpan (10—13 cm) just adjoining the cultivated layer the amount increased slightly 50 days after submerging, whereas for manganous manganese the amount increased even in deeper horizons in the latter period of submergence. In some paddy profiles the fluctuation of depth in amount of reduced iron and manganese is greater than that shown in Fig. 4–10.

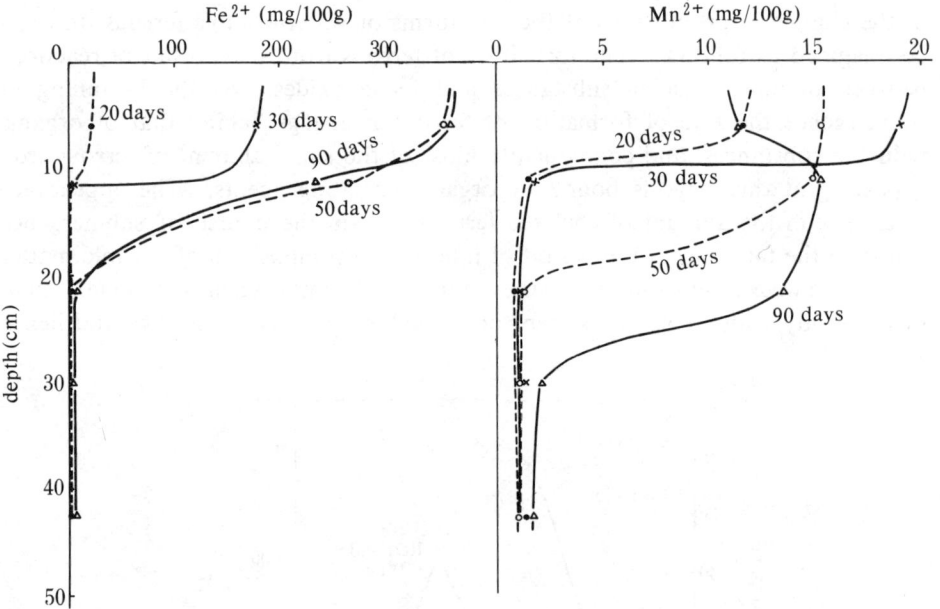

Fig. 4–10 Dynamics of iron and manganese in a paddy soil profile during the growing period of rice (Jiangsu, "White soil")[5]

The significance of changes of iron and manganese in paddy profiles in soil genesis will be discussed in Chapter 9.

REFERENCES

(1) Institute of Soil Science, 1961. Soil Environment of High–yield Rice. Chapter 6. Science Press, Beijing.

4.4 Dynamics of iron and manganese in paddy soils

(2) Xu Qi et al., 1980. The Paddy Soils of the Taihu Region in China. Science and Technology Press, Shanghai.
(3) Ding Chang-pu and Yu Tian-ren, 1958. Studies on oxidation-reduction processes in paddy soils. IV. Activities of iron and manganese in paddy soils derived from Red soil. Acta Pedologica, 6: 99–107.
(4) Yu Tian-ren, Ling Yun-xiao, Mu Ren-sheng and Liu Wan-lan, 1958. Effect of soil acidity on the activity of manganese. Soils Bulletin, 33: 16–30.
(5) Yu Tian-ren et al., 1959. Formation and melioration of low-yield "White soil" in the Taihu region. Acta Pedologica, 7: 42–58.
(6) Bao Xie-ming and Yu Tian-ren, 1964. Studies on oxidation-reduction processes in paddy soils. VI. Determination of complexed ferrous iron. Acta Pedologica, 12: 216–221.
(7) Bao Xie-ming, Liu Zhi-guang, Wu Jun and Yu Tian-ren, 1964. Studies on oxidation-reduction processes in paddy soils. VII. Forms of ferrous iron. Acta Pedologica, 12: 297–306.
(8) Cao Sheng-geng, 1964. Characteristics of the genesis of paddy soils derived from Red soil in the Jiangxi region. Acta Pedologica, 12: 155–163.
(9) Bao Xie-ming and Yu Tian-ren, 1978. Studies on oxidation-reduction processes in paddy soils. VIII. Characterization of water-soluble ferrous iron. Acta Pedologica, 15: 13–21.
(10) Bao Xie-ming, Liu Zhi-guang and Yu Tian-ren, 1978. Studies on oxidation-reduction processes in paddy soils. IX. Forms of water-soluble ferrous iron. Acta Pedologica, 15: 174–181.
(11) Bao Xie-ming, Ding Chang-pu and Yu Tian-ren, 1983. Stability constants of Mn(II)-complexes in soils as determined by the voltammetric method. Z. Pflanzenernähr. Bodenkunde, 146: 285–294.

CHAPTER 5

SULFUR

PAN SHU-ZHENG

Among inorganic oxidation–reduction systems sulfur differs from oxygen, iron and manganese discussed in the last chapters in that its transformation is not only controlled by physico–chemical laws but also accompanied by the participation of microorganisms in some processes. Hence in its transformation there are chemical oxidation–reduction reactions as well as biological oxidation–reduction processes. And, after the reduction of sulfur compounds the sulfide ions produced can further react with some metal ions. These characteristics of the behavior of the sulfur system add to the complexity of the physico–chemical properties of sulfur in paddy soils.

In this chapter, in addition to the amount, form and transformation of sulfur in paddy soils, special attention is paid to the chemical equilibria of sulfides.

5.1 REDUCTION OF SULFATE IN SOILS

The reduction of sulfate only occurs under strongly reducing conditions. This reduction is a biological process in nature, but is controlled by physico–chemical conditions. Hence it is related directly to the reducing strength of the medium and indirectly to the organic matter status of the soil, and is also influenced by environmental factors such as temperature. These will be discussed separately.

5.1.1 Effect of Eh and pH

The oxidation–reduction reaction of the sulfur system is:

$$SO_4^{2-} + 10H^+ + 8e \rightleftarrows H_2S + 4H_2O \tag{5-1}$$

At equilibrium:

$$pe = 5.12 - 1/8 \; pSO_4^{2-} + 1/8 \; pH_2S - 5/4 \; pH \tag{5-2}$$

$$Eh = 86.8 - 2.12 \; pSO_4^{2-} - 2.12 \; pH_2S - 21.2 \; pH \tag{5-3}$$

It is seen from the equation that at a constant pH the reduction of sulfate can occur only at low oxidation–reduction potential. It can be calculated that at a sulfate concentration of $10^{-2} M$ and a pH of 7 only at a pe of -3.13 (corresponding to -53 mV) can the concentration of H_2S attain a value of $10^{-6} M$, i.e., 0.034 ppm.

A practical measurement (Fig. 5–1) showed that after the submerging of the soil the concentration of H_2S and S^{2-} increased gradually with the fall of the

5.1 Reduction of sulfate in soils

Eh. At the sixth day the pH_2S and pS^{2-} values[1] were 6.8 and 15.2 respectively. At this time the *Eh* was -30 mV and when corrected to Eh_7 was -95 mV, close to the theoretically calculated value.

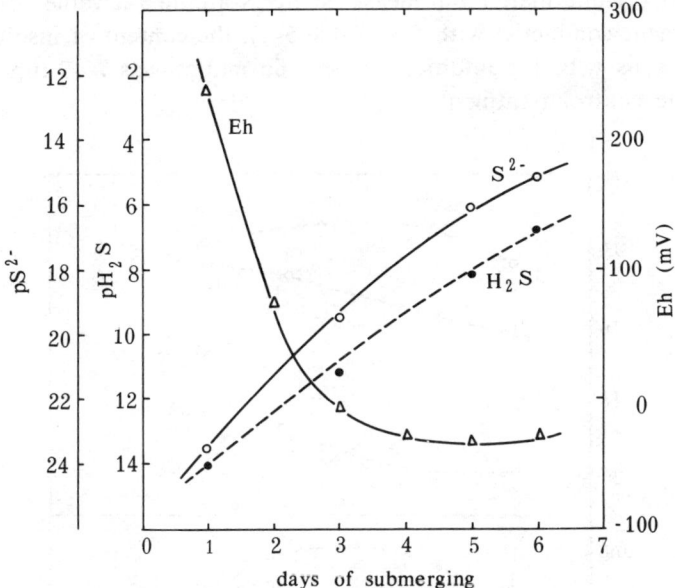

Fig. 5–1 Changes in H_2S and S^{2-} contents of the soil after submerging in relation to *Eh* (paddy soil derived from red soil)

Regarding the effect of pH, it can be assumed from physico–chemical considerations that a low pH should favor the process of sulfate reduction, for Equation (5–1) shows that in the formation of H_2S there is consumption of H^+ ions. However, because the reduction of SO_4^{2-} is accomplished through biological activities of anaerobic bacteria, a pH low enough to retard the activity of these bacteria will lead to a low rate of SO_4^{2-} reduction. For instance, in an acid sulfate soil with a pH of 2 the pH_2S value was still larger than 10 one month after submerging, whereas in a sandy paddy soil with a pH of about 7 the pH_2S value dropped to about 6 only one day after submerging if organic matter and K_2SO_4 had been added to the soil.

The effect of pH on the chemical equilibrium of sulfides will be discussed in a latter section.

5.1.2 Effect of organic matter

In addition to its sulfur component, organic matter is the principal source of electrons necessary for the creation of a strongly reducing condition of the soil. Therefore, it can affect the reduction of sulfate markedly. An experiment showed that for a red soil containing very little organic matter the sulfide

---

[1] The pH_2S and pS^{2-} cited in this chapter were determined by a H_2S-gas sensor and a sulfide selective electrode respectively.

content was very low even when it was submerged at a high temperature (Fig. 5-2). On the other hand, the S^{2-} content may be higher by several orders of magnitude if organic matter has been added. In the treatment with an added large amount of organic matter the measured pH_2S attained a value of 5.1. In another experiment conducted with S^{35} (Table 5-1), the content of insoluble sulfides in paddy soils with the addition of organic matter was 7—9 times higher than that of the control treatment.

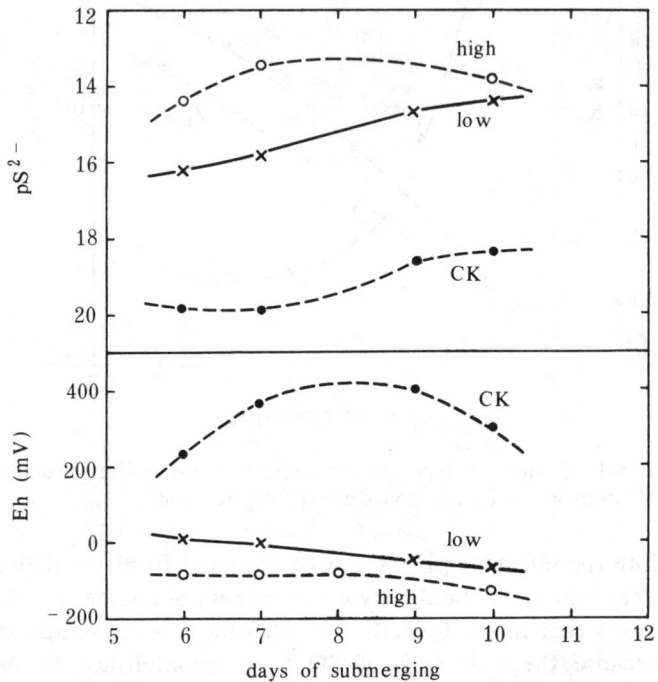

Fig. 5-2 Effect of organic matter (green manure) on S^{2-} formation in a submerged red soil (low and high correspond to 20 and 40 ton/ha respectively)

TABLE 5-1 Effect of organic matter on sulfate reduction in submerged paddy soils[1]

Soil type	Treatment*	Sulfides (mg/100g)		
		3rd day	10th day	20th day
Acid	CK	0.55	0.61	0.76
	2% wheat straw	1.46	5.35	2.87
Neutral	CK	0.57	0.62	1.98
	2% wheat straw	0.73	5.20	3.85
Calcareous	CK	0.63	0.58	0.73
	2% wheat straw	0.68	5.08	7.17

* 0.1% $(NH_4)_2SO_4$ added

5.1.3 Effect of temperature

Temperature affects the reduction of sulfate through influencing the activities of microorganisms. It can be seen from Table 5-2 that for a sandy paddy soil with an added 5% of organic matter the contents of H_2S and S^{2-} increased with the increase in temperature of incubation. The difference in content of H_2S may be as high as four orders of magnitude between those incubated at 30°C and at 0°C.

TABLE 5-2 Influence of temperature on reduction of sulfate

Temperature (°C)	Eh (mV)	pH	pS^{2-}	pH_2S
0	−99	7.0	15.6	8.7
10	−164	6.7	15.2	6.6
30	−189	6.5	14.1	4.8

5.2 FACTORS AFFECTING THE CHEMICAL EQUILIBRIA OF SULFIDES IN SOILS

After the reduction of sulfate to sulfide, the proportions of molecular hydrogen sulfide, insoluble sulfide and soluble sulfide ions are determined by the chemical equilibria among them. In this regard the association–dissociation between S^{2-} and H^+ ions and the precipitation–solution between S^{2-} and metal ions will play a dominant role. Therefore, the amount of sulfide ion itself and such factors as pH, iron, manganese and zinc ions which can form precipitates with S^{2-} can all affect these chemical equilibria. These factors will be discussed in order.

5.2.1 pH

The dissociation of H_2S proceeds in two steps:
$H_2S \rightleftarrows HS^- + H^+$,

$$K_1 = \frac{(H^+)(HS^-)}{(H_2S)} = 1.6 \times 10^{-7} \tag{5-4}$$

$HS^- \rightleftarrows S^{2-} + H^+$,

$$K_2 = \frac{(H^+)(S^{2-})}{(HS^-)} = 7.9 \times 10^{-15} \tag{5-5}$$

$$(S^{2-}) = K_1 \cdot K_2 \cdot \frac{(H_2S)}{(H^+)^2} \tag{5-6}$$

Inserting the K values in Equation (5-6) and taking logarithmic form, we get:

$$pH_2S = 2pH + pS^{2-} - 20.9 \tag{5-7}$$

$$pS^{2-} = pH_2S + 20.9 - 2pH \tag{5-8}$$

It is seen from the equations that the pH of the medium plays a dominant role in governing the dissociation of H_2S. Under conditions where the quantity of sulfide is kept constant the quantity of S^{2-} will increase and that of H_2S decrease with the increase in pH. However, since the pH of the medium will also affect the solubilities of sulfides of iron, manganese, etc., in practice the relationship between pH and the chemical equilibria of sulfides in paddy soils is more complicated.

In an experiment three submerged paddy soils were adjusted to different pH values, and then concentrations of H_2S and S^{2-} were determined. The results are shown in Figs. 5-3, 5-4 and 5-5.

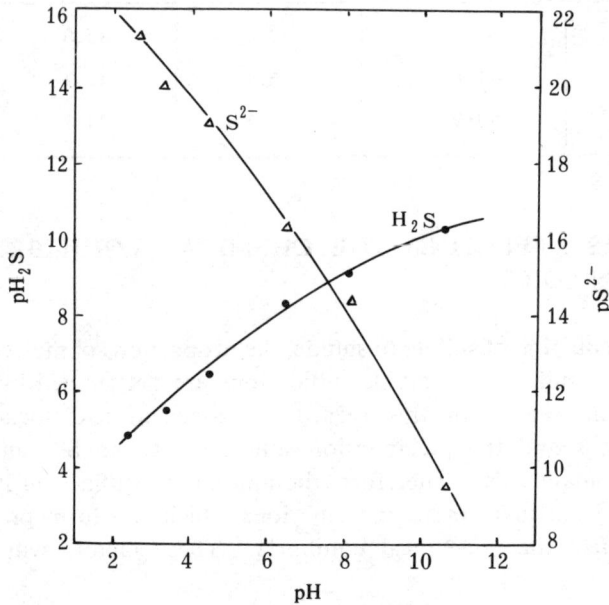

Fig. 5-3 pH_2S and pS^{2-} in relation to pH for a paddy soil derived from red soil[6]

The results show that in spite of differences in the absolute amount of sulfide and other factors in the three soils, a tendency toward an increase in H_2S concentration with a fall of pH is general. In the pH range of 4—8 the slopes of the pH_2S–pH lines for the three soils are 0.8, 0.9 and 1.1 respectively. It should be noted that the dependence of pS^{2-} on the pH is also nearly linear, with a change of pS^{2-} of 1.1—1.5 units per one unit change in the pH.

The slopes of pH_2S–pH and pS^{2-}–pH lines for these three soils are all inclined more than those calculated from Equations (5-7) and (5-8). This may serve as a proof of the involvement of the precipitation–solution equilibria of FeS, etc. along with the association–dissociation equilibria of Equations (5-4) and (5-5).

The pH_2S in a submerged paddy soil derived from red soil and in a submerged

5.2 Factors affecting the chemical equilibria of sulfides in soils

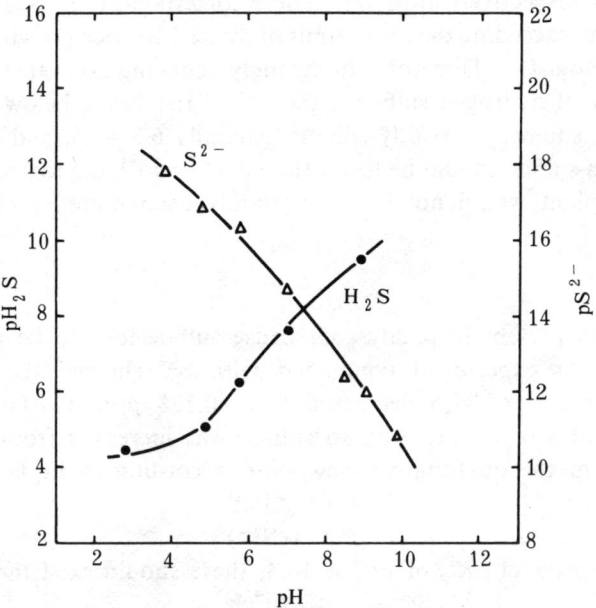

Fig. 5-4 pH$_2$S and pS^{2-} in relation to pH for a paddy soil derived from red soil plus organic matter[5]

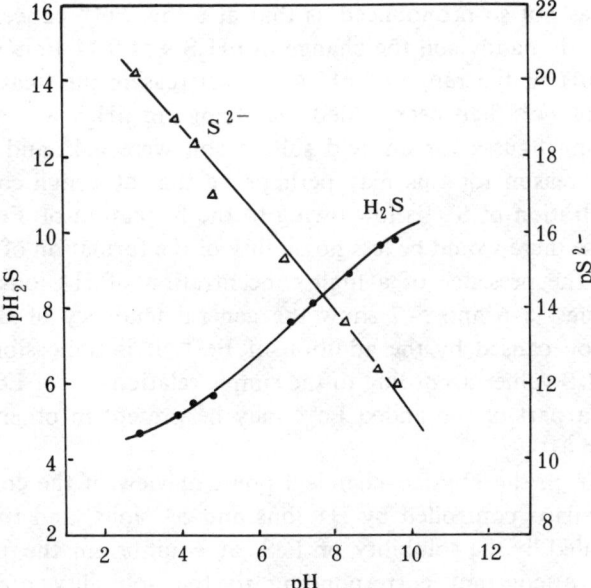

Fig. 5-5 pH$_2$S and pS^{2-} in relation to pH for a paddy soil derived from lacustrine deposit[5]

paddy soil derived from lacustrine deposit after the addition of organic matter were 5.2 and 5.7 respectively at pH 5. These correspond to a H_2S concentration of 0.2—0.07 ppm, exceeding the lower limit of toxicity for rice growth as established by plant physiologists. Therefore, in strongly reducing soils there is the possibility of toxicity of hydrogen sulfide if the pH of the soil is below 5. However, since the pH of submerged paddy soils is generally 6.5—7.5, and the concentration of hydrogen sulfide should be lower than 0.03 ppm[3], the toxicity of hydrogen sulfide to rice plant would not be a commonly encountered problem.

5.2.2 Fe^{2+}

Ferrous ions present in paddy soils cause sulfide ions to be precipitated as insoluble FeS. An experiment conducted with S^{35} showed that at a pH of 6.0 the concentration of H_2S decreased from 0.135 ppm to 0.019 ppm when the molecular ratio of ferrous iron to sulfide was increased from 1:1 to 3:1[2]. Considering from the quantitative viewpoint, according to Equation (5–10):

$$Fe^{2+} + S^{2-} \rightleftarrows FeS \tag{5-9}$$
$$K_{sp} = (Fe^{2+})(S^{2-}) \tag{5-10}$$

if the solubility product pK_{sp} of FeS is 18.4, there should exist the relationship:

$$pS^{2-} = 18.4 - pFe^{2+} \tag{5-11}$$

In order to elucidate the effect of ferrous iron on the chemical equilibrium of sulfides, an experiment was carried out on two reduced paddy soils. Various amounts of Fe^{2+} were added to the soil, and the concentration of H_2S was determined at different pH levels of the soil. The results shown in Figs. 5–6 and 5–7 revealed that the addition of Fe^{2+} led to a decrease in the concentration of H_2S. It is worthy of note that the effect of pH on pH_2S at a high Fe^{2+} concentration was not so pronounced as that at a low Fe^{2+} concentration. For example, in a sandy paddy soil the change in pH_2S was 0.44 unit for the change of one unit of pH in the range of pH 4—6, whereas in the treatment to which a high amount of Fe^{2+} had been added the change in pH_2S was only 0.27 unit. The corresponding figures for an acid sulfate soil were 0.45 and 0.11 unit respectively. The reason for this may perhaps be that at a high concentration of Fe^{2+} the concentration of S^{2-} is low owing to the formation of FeS precipitates, and in such a case there would be less possibility of the formation of large amounts of H_2S even in the presence of a high concentration of H^+ ions.

Although Figs. 5–6 and 5–7 show the general tendency of the decrease of H_2S concentration caused by the addition of Fe^{2+}, it is impossible to calculate the pS^{2-} and pH_2S values according to the simple relationship of Equation (5–11). This is because a part of the added Fe^{2+} may be present in other forms in the soil (cf. Chapter 4).

Reasoning from the physico–chemical point of view, if the concentration of H_2S in paddy soils is controlled by H^+ ions and S^{2-} ions, and the latter factor is in turn controlled by the solubility of FeS, at equilibrium the product $(Fe^{2+})(S^{2-})$ should be a constant corresponding to the solubility product of FeS. Taking the pK_{sp} value as 18.4, the concentration of S^{2-} at a water–soluble ferrous iron concentration of 23 ppm should be $10^{-15}M$, namely 3.2×10^{-11} ppm.

5.2 Factors affecting the chemical equilibria of sulfides in soils

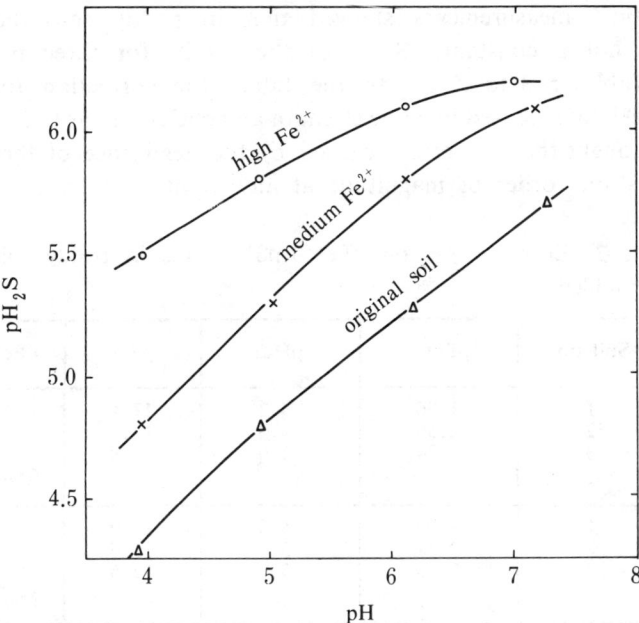

Fig. 5-6 Effect of Fe^{2+} on chemical equilibria of sulfides for a loamy paddy soil[5] (low and high correspond to 20 and 40 m mole/100g respectively)

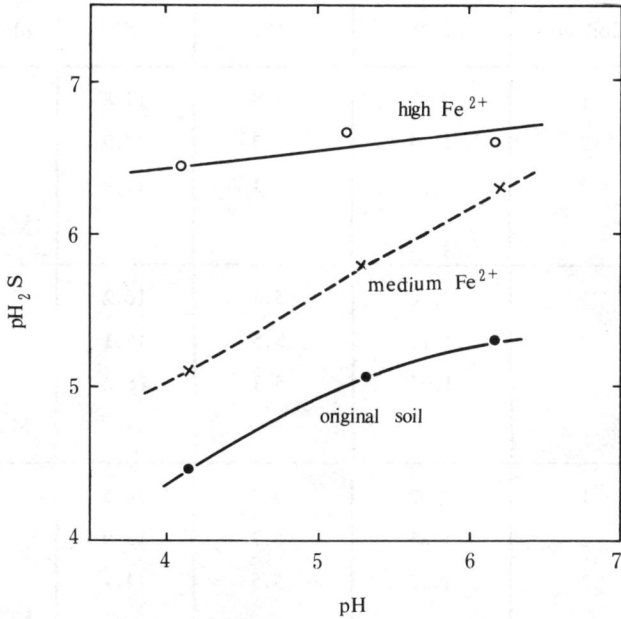

Fig. 5-7 Effect of Fe^{2+} on chemical equilibria of sulfides for an acid sulfate paddy soil[5]

However, practical measurements showed that in paddy soils the product $(Fe^{2+})(S^{2-})$ was not a constant. Some of the results for three paddy soils are shown in Tables 5-3 to 5-5. In the tables the correction for activity coefficient of Fe^{2+} ions caused by ionic strength and chelation has been neglected. However, it is thought that the errors caused by the negligence of these factors would not exceed one order of magnitude at any event.

TABLE 5-3 Effect of pH on $pFe^{2+}+pS^{2-}$ value in paddy soils (no Fe^{2+} added)[5]

pH	Soil no.	pFe^{2+}	pH_2S	pS^{2-}	$pFe^{2+}+pS^{2-}$
3.9—4.1	1 2 3	1.96 2.50 2.24	4.3 4.5 5.0	17.4 17.2 17.9	19.4 19.7 20.1 Mean 19.7
4.9—5.2	1 2 3	2.09 2.37 3.00	4.8 5.1 5.0	15.9 16.0 15.6	18.0 18.4 18.6 Mean 18.3
6.2	1	2.68	5.3	13.9	16.6

TABLE 5-4 Effect of pH on $pFe^{2+}+pS^{2-}$ value in paddy soils (medium amount of Fe^{2+} added)[5]

pH	Soil no.	pFe^{2+}	pH_2S	pS^{2-}	$pFe^{2+}+pS^{2-}$
3.5—4.1	1	0.82	4.8	17.8	18.6
	2	1.04	5.3	17.9	18.9
	3	1.07	5.1	17.9	19.0
					Mean 18.8
4.9—5.2	1	0.86	5.3	16.2	17.1
	2	1.11	5.5	16.1	17.2
	3	1.07	5.3	16.2	17.3
					Mean 17.2
6.1—6.4	1	0.92	5.8	14.5	15.4
	2	1.85	5.7	13.9	15.7
	3	1.30	5.5	14.1	15.4
					Mean 15.5
7.2	1	1.45	6.1	12.7	14.2

5.2 Factors affecting the chemical equilibria of sulfides in soils

TABLE 5–5 Effect of pH on $pFe^{2+}+pS^{2-}$ value in paddy soils (high amount of Fe^{2+} added)[5]

pH	Soil no.	pFe^{2+}	pH_2S	pS^{2-}	$pFe^{2+}+pS^{2-}$
3.9—4.1	1	0.59	5.5	18.5	19.1
	2	0.80	5.3	18.0	18.8
	3	0.47	5.3	17.9	18.4
					Mean 18.8
4.9—5.2	1	0.59	5.8	16.9	17.5
	2	0.85	5.6	16.1	17.0
	3	0.51	5.7	16.3	16.8
					Mean 17.1
6.1—6.4	1	0.60	6.1	14.8	15.4
	2	1.44	5.8	13.9	15.4
	3	0.77	5.9	14.5	15.3
					Mean 15.4
6.9—7.3	1	0.86	6.2	13.1	14.1
	3	1.41	6.5	13.7	15.1
					Mean 14.6

It can be seen from the tables that the $pFe^{2+}+pS^{2-}$ value is affected by pH, and in the pH range of 4—7 the numeral values may differ by 4.5 units. For original soils without the addition of Fe^{2+} the $pFe^{2+}+pS^{2-}$ value was close to the theoretical value of 18.4 at about pH 5, whereas for treatments with the addition of Fe^{2+} the value approximated to the theoretical value at about pH 4. For all treatments the $pFe^{2+}+pS^{2-}$ value was smaller than 18.4 at high pH, and the higher the pH the more pronounced this effect. It will also be noticed from the comparison of the three tables that at the same pH the $pFe^{2+}+pS^{2-}$ value of the soil with an added Fe^{2+} was smaller than that of the soil into which Fe^{2+} was not added.

In Fig. 5–8 is shown the dynamic change in the $pFe^{2+}+pS^{2-}$ value in the course of the submergence of paddy soils. The overall change may amount to four units. The minimum appeared at the 4th—5th day after submergence. The $pFe^{2+}+pS^{2-}$ curve showed a parallelism with the pFe^{2+} curve in which there also appeared a minimum at the 4th—5th day. This may serve as another explanation for the supposition that the wide range fluctuation of Fe^{2+} concentration is an important factor for the inconstancy of $pFe^{2+}+pS^{2-}$ values in soils.

It is not yet clear what the real cause is of the change of the water–soluble $pFe^{2+}+pS^{2-}$ value with soil conditions. Possibly this may be related to the formation of precipitates of other forms of Fe^{2+} such as $Fe(OH)_2$ and other forms of S^{2-} such as FeS_2. An additional evidence for this supposition is the smaller effect of the Fe^{2+} addition on the $pFe^{2+}+pS^{2-}$ value at higher pH for the two soils.

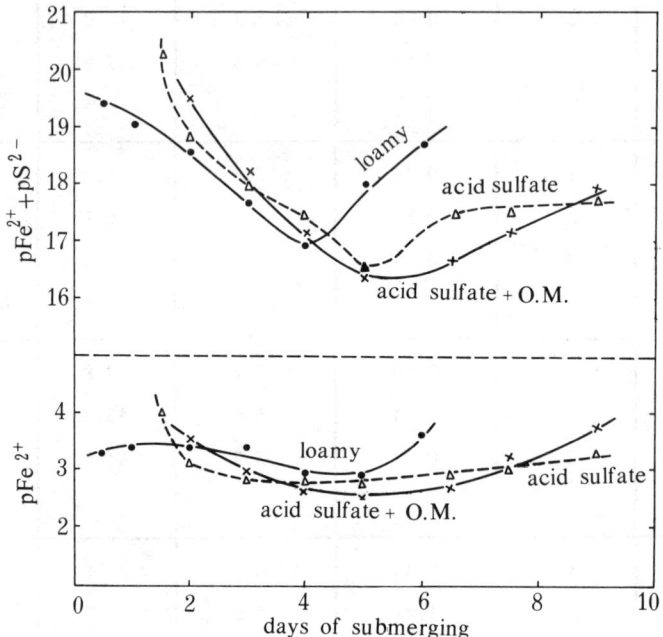

Fig. 5-8 Change in $pFe^{2+}+pS^{2-}$ during submergence of paddy soils

5.2.3 Mn^{2+} and Zn^{2+}

The effect of manganous ions on the chemical equilibria of sulfides is similar to that of ferrous iron. It can be seen from Figs. 5–9 and 5–10 for two paddy soils with added organic matter that the addition of Mn^{2+} can also cause the decrease in H_2S concentration. The difference between Mn^{2+} and Fe^{2+} is that the effect of the former is more pronounced at higher pH, whereas that of the latter is more pronounced at a lower pH. This seems to be related to the difference in the change of solubilities of other precipitates of these elements (such as hydroxides) with the change in pH.

The solubility product of ZnS (pK_{sp} 22—25) is much smaller than that of FeS or MnS. Therefore, if Zn^{2+} ions are added to the soil, there will be an immediate precipitation of ZnS, causing a sudden decrease in H_2S concentration. For a submerged loamy paddy soil the pH_2S value increased from 5.3 to 7.0 at pH 5.5 after the addition of Zn^{2+}, and for a submerged acid sulfate soil the addition of Zn^{2+} caused the H_2S concentration to decrease by 4.5 orders of magnitude. These data suggest that the effect of Zn^{2+} on the concentration of soluble sulfides is much more pronounced than that of Fe^{2+} and Mn^{2+}.

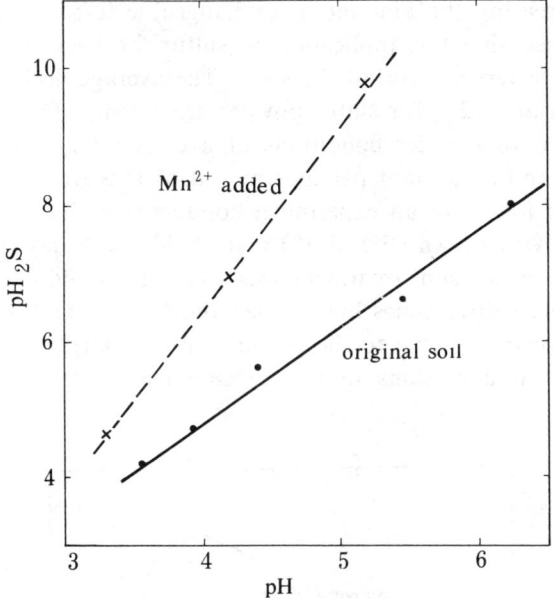

Fig. 5-9 Effect of Mn^{2+} on chemical equilibria of sulfides for an acid sulfate paddy soil (8 m mole Mn^{2+}/100g added)

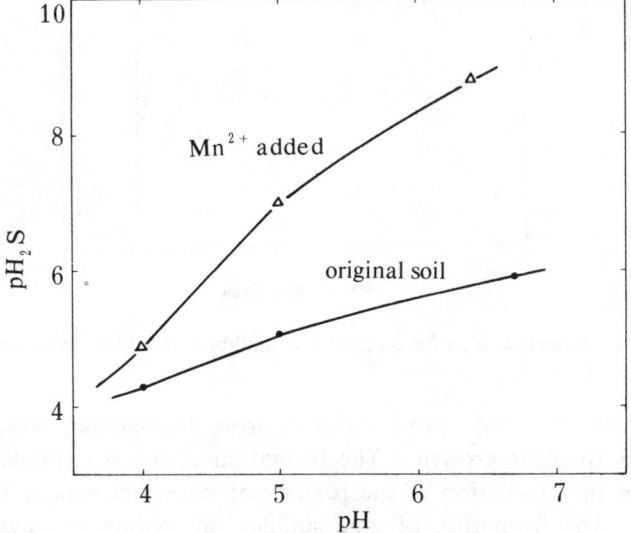

Fig. 5-10 Effect of Mn^{2+} on chemical equilibria of sulfides for a loamy paddy soil

5.2.4 Amount of sulfide

The data in Table 4-9 of Chapter 4 have shown that sulfide ions can compete for ferrous iron with cation-exchange sites of the soil to form ferrous sulfide precipitates, thus decreasing the amount of exchangeable ferrous iron. A field experiment[4] confirmed that the application of sulfur fertilizers led to the decrease in the content of ferrous iron of the soil. The average decrease was 17% for gypsum treatment and 12% for sulfur powder treatment. On the contrary, it should be suspected that under conditions of a constant amount of ferrous and manganous ions and a constant pH the amount of H_2S would increase with the increase in sulfide ions. In an experiment conducted with a red soil added with green manure it was shown (Fig. 5-11) that at the ninth day after submergence the pH_2S value in gypsum treatment was 4.6, corresponding to 0.85 ppm of H_2S which was about three times higher than the H_2S content in the check treatment. This is apparently due to the production of a larger amount of sulfide ions under reduced conditions in the gypsum treatment.

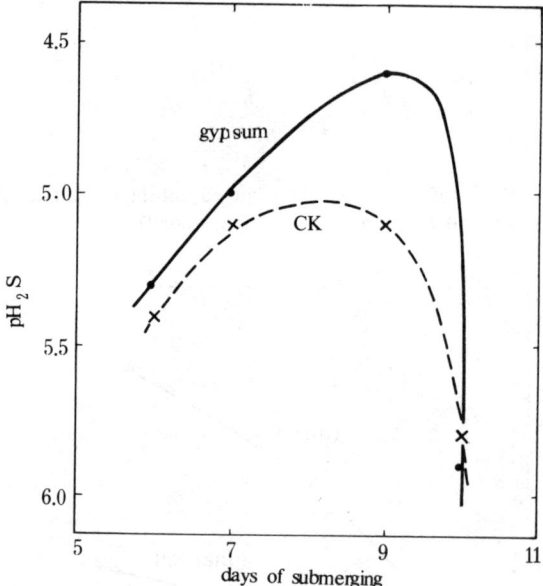

Fig. 5-11 Influence of sulfur on chemical equilibria of sulfides in reduced soil

The formation of sulfide precipitates of iron, manganese, zinc, etc. is of great significance in plant growth. The formation of ferrous sulfides is an important factor in the alleviation of the toxicity of large amounts of ferrous iron to rice growth. The formation of zinc sulfides can reduce the availability of zinc for plants, especially when the pH of the soil is high. As to the behavior of sulfur itself, since in ordinary paddy soils there is always the production of a certain amount of sulfide ions under reduced conditions, it is only owing to the formation of insoluble metal sulfides by most of the sulfide that the toxicity of hydrogen sulfide to rice growth is not a commonly encountered problem. In reduced paddy soils the probability and the extent of toxicity of hydrogen sulfide to plant growth are primarily dependent on the position of the chemical equilibrium among various forms of sulfides determined by the amount of sulfide ions, the contents of certain metal ions, especially ferrous iron, and environmental conditions, particularly the pH of the soil.

5.3 DYNAMICS OF SULFIDES IN PADDY SOILS

In the course of submergence of the soil, the dynamics of sulfides is controlled by the biological reduction of sulfate as well as by a series of chemical properties of the soil. In the following, three soils are taken as examples for discussion.

5.3.1 Acid sulfate soil

Acid sulfate soils are characterized by a high content of sulfur (0.4—3% S) and a very low pH when they are dried. The pH of the soil also remains at a low level within a certain period of submergence. In an experiment a reclaimed acid sulfate soil was submerged with and without the addition of organic matter, and changes in the pH_2S, pH and Fe^{2+} concentration were recorded. The results are shown in Figs. 5–12 and 5–13.

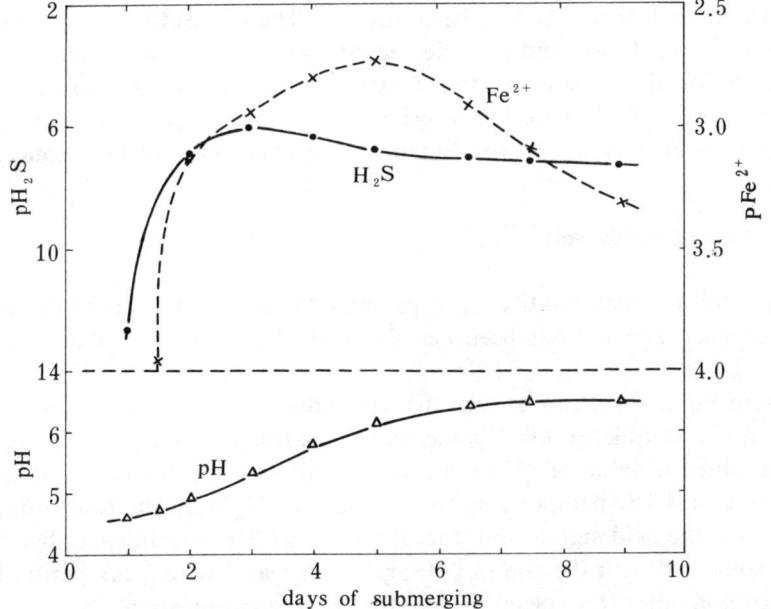

Fig. 5–12 Changes in pH_2S, pH and Fe^{2+} concentration during submergence for an acid sulfate paddy soil

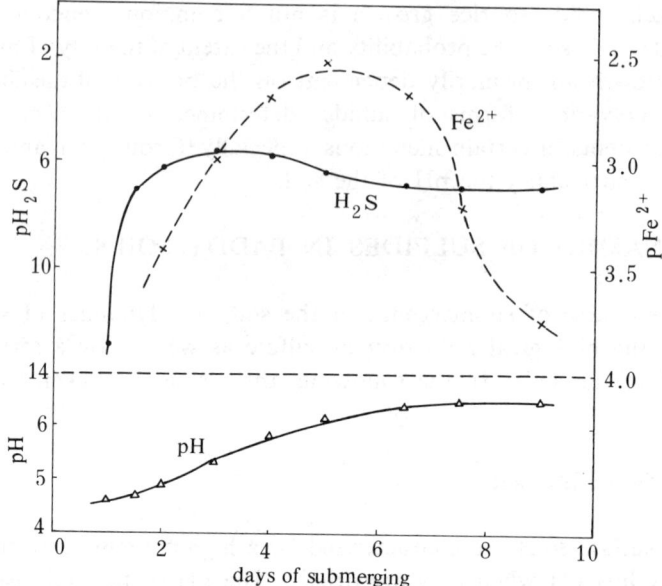

Fig. 5-13 Changes in pH_2S, pH and Fe^{2+} concentration during submergence for an acid sulfate paddy soil plus organic matter[5]

The H_2S concentration increased rapidly after one day's submergence, especially when organic matter had been added. The concentration reached a peak value on the third day, and then decreased again. The peak value of H_2S concentration for the organic matter treatment was pH_2S 5.6, corresponding to 0.085 ppm of H_2S. The decline in H_2S concentration after the peak value was parallel with an increase of pH but not with a decrease in Fe^{2+} concentration.

5.3.2 Loamy paddy soil

This soil is representative in large areas of the granite mountain regions of south China. The soil has been heavily leached, and has a low content of iron oxides.

From Fig. 5-14 it can be seen that the reduction of sulfate proceeded rapidly owing to the abundance of organic matter and high fertility of the soil. The pH_2S attained a value of about 6 after one day's submergence, and reached a peak value of 4.7 (corresponding to 0.68 ppm of H_2S) on the fourth day. Like the case for the acid sulfate soil, the decline in H_2S concentration after the peak period coincided with the rise in pH, and there was also a peak period for Fe^{2+} concentration after the peak period for H_2S concentration.

If a comparison is made between Fig. 5-14 and Figs. 5-12 and 5-13, it can be observed that the amount of water-soluble Fe^{2+} in the loamy paddy soil was

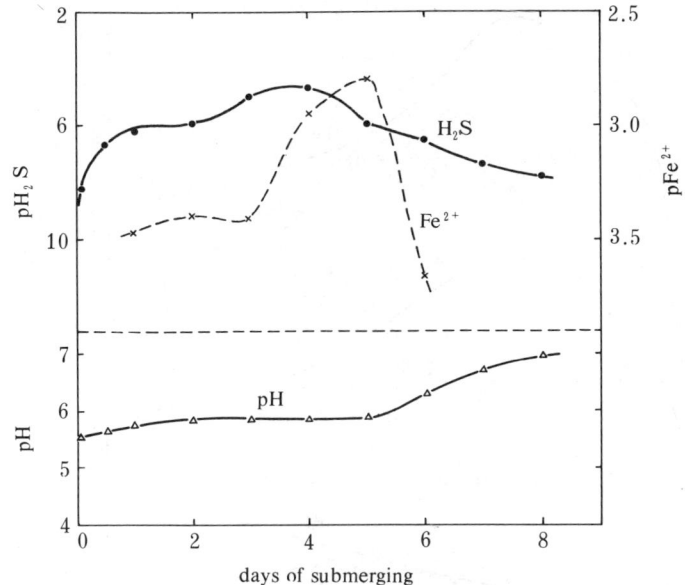

Fig. 5–14 Changes in pH$_2$S, pH and Fe^{2+} concentration during submergence for a loamy paddy soil (organic matter and K$_2$SO$_4$ added)[5]

generally lower than that in the acid sulfate soil. It may be suspected that if the loamy soil is submerged at a high temperature after the addition of large amounts of easily decomposable organic matter, the possibility of a H$_2$S concentration high enough for the toxicity to rice growth should exist.

5.3.3 Paddy soil derived from red soil

This soil is derived from Quaternary red clay and has a high content of iron oxides. Since the fertility level is low and microbiological activity is weak, the H$_2$S concentration increases only slowly after submerging, attaining a peak value on the ninth day (Fig. 5–15). The peak H$_2$S concentration (0.007 ppm, corresponding to pH$_2$S 6.7) was much lower than that of the two above–mentioned soils. This may perhaps be related to the high content of Fe^{2+} of this soil. It can be seen from the figure that the Fe^{2+} concentration on the third day of submergence exceeded the maximum Fe^{2+} concentration of the two above–mentioned soils, and maintained at a high level within a certain period. This high Fe^{2+} concentration should affect the chemical equilibria of sulfides, lowering the concentration of water–soluble sulfide. It can also be seen from the figure that for this soil the pFe^{2+} curve paralleled the pH$_2$S curve quite well.

The dynamics of H$_2$S concentration for the three soils described above may serve to explain the regularity of the chemical equilibria of sulfides discussed in section 5.2, i.e., the concentration of H$_2$S in submerged paddy soils is strongly affected by the Fe^{2+} concentration and the pH of the soil, along with the amount of sulfide.

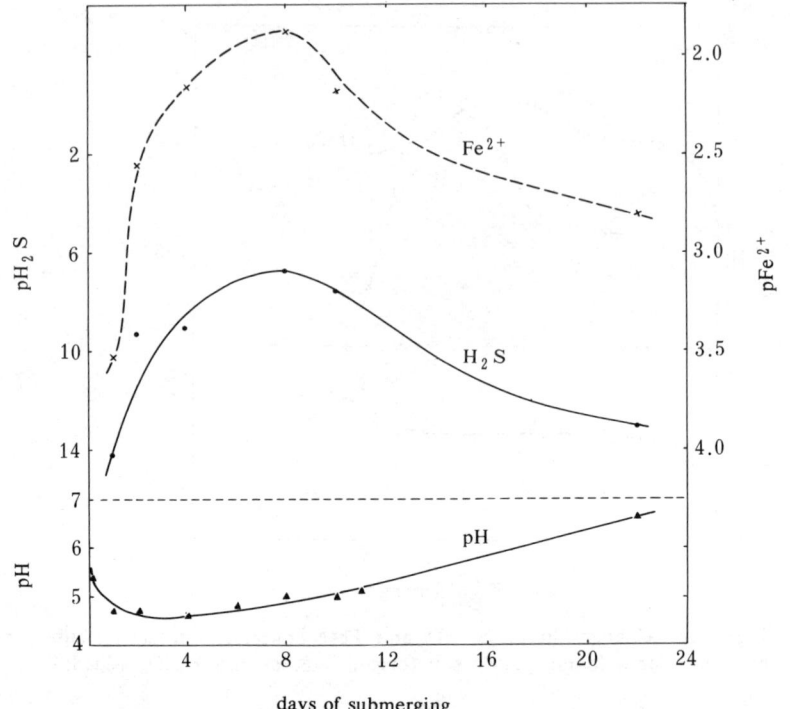

Fig. 5-15 Changes in pH_2S, pH and Fe^{2+} concentration during submergence for a paddy soil derived from red soil (organic matter and K_2SO_4 added)

5.4 FORMS AND AMOUNT OF SULFUR IN PADDY SOILS

Sulfur compounds in paddy soils include both organic and inorganic parts. The former accounts for 86—94% of the total sulfur. In Table 5-6 is shown the sulfur content of paddy soils of South China derived from different parent materials. It is seen that except for those derived from recent river alluvium

TABLE 5-6 Sulfur content of the cultivated layer of paddy soils of South China[4]

Parent material	Sulfur		
	Total (S%)	Organic (S%)	Available (ppm S)
Granite	0.027	0.025	16.2
Quaternary red clay	0.027	0.025	18.9
Sedimentary rocks	0.027	0.025	22.8
Recent river alluvium	0.016	0.014	14.6
Lacustrine deposit	0.028	0.024	38.6
Mean	0.025	0.023	22.2

5.4 Forms and amount of sulfur in paddy soils

in which the sulfur content is relatively low, there is no significant difference in sulfur content among various parent materials. However, the acid sulfate paddy soil along seashores may have a sulfur content as high as 3%.

Under oxidized conditions more than 90% of the available sulfur is in the form of sulfate. Under strongly reducing conditions the oxidized form of sulfur is reduced to sulfide. The amount of sulfide in soils is determined by the amount of reducible sulfur compounds as well as by the amount of organic reducing substances formed under submerged conditions, and this is the reason the content of sulfide is high in acid sulfate soils containing large amounts of sulfate and organic matter. In some paddy soils derived from lacustrine deposit the content of sulfide is also high. For instance, in the cultivated layer of a swampy paddy soil the sulfide content was 39 mg per 100 g of soil, corresponding to 1.2 millimole[2].

The amounts of various forms of sulfide in submerged soils are related not only to the reducing condition, but also to other factors particularly the pH of the soil. For example, it was shown in an experiment that if the soil was submerged at a high temperature after the addition of large amounts of easily

TABLE 5-7 pS^{2-} and pH_2S values of some submerged paddy soils*

Soil	Locality	Eh (mV)	pH	pS^{2-}	pH_2S
Red soil	Jinxian	−94	5.8	13.9	4.2
Loamy	Zixi	−226	6.8	13.2	5.2
Acid sulfate	Lianjiang	−181	6.3	13.2	4.7
Muddy	Shunchang	−281	6.9	12.5	5.4 (calc.)

* Organic matter and K_2SO_4 added, incubated at 38°C

TABLE 5-8 H_2S and S^{2-} contents of some paddy soils derived from red soil (in situ) (Jinxian) (October)[5]

Soil type	Reduction intensity	pS^{2-}	pH_2S
Old paddy soil	Strong	14.6	6.6
	Moderate	20.7	9.5
	Weak	20.7	11.1
Young paddy soil	Strong	14.6	6.6
	Moderate	14.6	6.5
	Weak	18.4	11.4
Alluvial	Moderate	14.3	7.0

decomposable organic matter, the content of molecular H_2S in the period when the pH was still not high enough might be so high as to be harmful to rice growth (Table 5-7). However, from field measurements shown in Tables 5-8 and 5-9 it can be seen that even in some strongly reducing paddy soils the pH_2S, under the steady decomposition of soil organic matter, was at a level higher than 6. The pS^{2-} for all paddy soils were higher than 14 (Tables 5-8 and 5-9). Only in local micro-regions where rice plants showed toxic symptoms in a newly-formed paddy field developed on red soil was the pH_2S value 4.7, corresponding to a H_2S concentration of 0.68 ppm.

TABLE 5-9 H_2S and S^{2-} contents in paddy soils of different regions (in situ) (July)

Locality	Soil	Eh (mV)	pH	pS^{2-}	pH_2S
Jinxian	Fertile	-7	7.2	15.6	9.1*
	Fertile	55	6.7	15.2	7.3
	Ordinary	86	7.4	15	8.9*
Shunchang	Ordinary	-32	6.2	15.4	6.9*
	Muddy	15	7.0	14.2	7.0
	Ordinary	208	6.3	16.1	7.8*
Shaxian	Muddy	-70	7.0	15.0	6.7
	Muddy	-50	6.2	14.6	6.5
	Ordinary	140	5.9	17.1	8.6

* Calculated according to Equation (5-7)

From what has been discussed above it may be concluded that the amount of water-soluble sulfide in paddy soils would not generally be so high as to be harmful to rice growth. This will be discussed further in Chapter 10.

REFERENCES

(1) Institute of Soil Science, 1961. Soil Environment of High-yield Rice. Chapter 6. Science Press, Beijing.
(2) Institute of Soil Science, 1978. Soils of China. Part B, Chapter 10. Science Press, Beijing.
(3) Yu Tian-ren and Liu Zhi-guang, 1964. Oxidation-reduction processes in paddy soils and their relationship to rice growth. Acta Pedologica, **12**: 380-389.
(4) Liu Chung-qun, Chen Gue-an, Cao Shu-qing and Liu Yuan-chang, 1981. Sulfur status and sulfur fertilization in soils of South China. Acta Pedologica, **18**: 185-198.
(5) Pan Shu-zheng, Liu Zhi-guang and Yu Tian-ren, 1982. Chemical equilibria of sulfides in submerged soils as studied with a H_2S-sensor. Soil Sci., **134**: 171-175.

CHAPTER 6

ION ADSORPTION

ZHANG XIAO-NIAN

Ion adsorption is one of the important physico-chemical properties of soil. The principal cause for ion adsorption is the electric charge carried by soil particles. Paddy soils are characterized by a low iron oxide content due to reductive eluviation caused by long-term submergence and a high content of organic matter as compared with their corresponding upland soils. These properties together with a periodical change in pH during alternate wetting and drying have some influence on the electric charge of the soil. The ionic composition of soil solutions during submergence also differs from that under upland conditions. These characteristics of paddy soil make the behaviors of ion adsorption different from upland soils in many respects.

In this chapter are first discussed the electric charges of paddy soils, the interactions between ions and soil particles, and finally, the peculiarily in composition of exchangeable cations in paddy soils.

6.1 ELECTRIC CHARGE OF PADDY SOILS

6.1.1 Property of electric charge

The electric charge of a soil is primarily determined by clay minerals and oxides of iron and aluminum, and is also related to the organic matter in the soil. During the development of a paddy soil there is frequently a remarkable movement and differentiation of iron and manganese, and also an alteration in the composition of clay minerals. Therefore, although the property of a paddy soil with respect to electric charge chiefly originates from its parent soil, quantitatively the electric charge of a paddy soil may differ considerably from its parent soil[7].

On the surface of soil particles there are both negative charges and positive charges, and the former may be further distinguished as permanent charges and variable charges. The algebraic sum of positive charges and negative charges is the net charge. The sign of the net charge is positive if the quantity of positive charge is larger than that of negative charge, and vice versa.

In China most of the paddy soils are distributed in the south. Paddy soils derived from red soils in South China are characterized by a large amount of positive charge and a high proportion of variable negative charge as compared with paddy soils in North China. For some paddy soils derived from laterites or lateritic soils there may even appear an iso-electric point. In the paddy soils of North China, the positive charge is generally small, and the permanent negative charge far exceeds the variable negative charge.

The quantity of electric charge of the soil usually changes with the change in pH. From the titration curves shown in Fig. 6-1 it can be seen that at a same pH the quantity of negative charge is smaller for a paddy soil derived from laterite than a paddy soil derived from yellow-brown soil, due chiefly to the difference in composition of clay minerals. The larger amount of negative charge in the paddy soil derived from laterite as compared with its parent soil is due probably to the loss of free iron oxides during the development of a paddy soil.

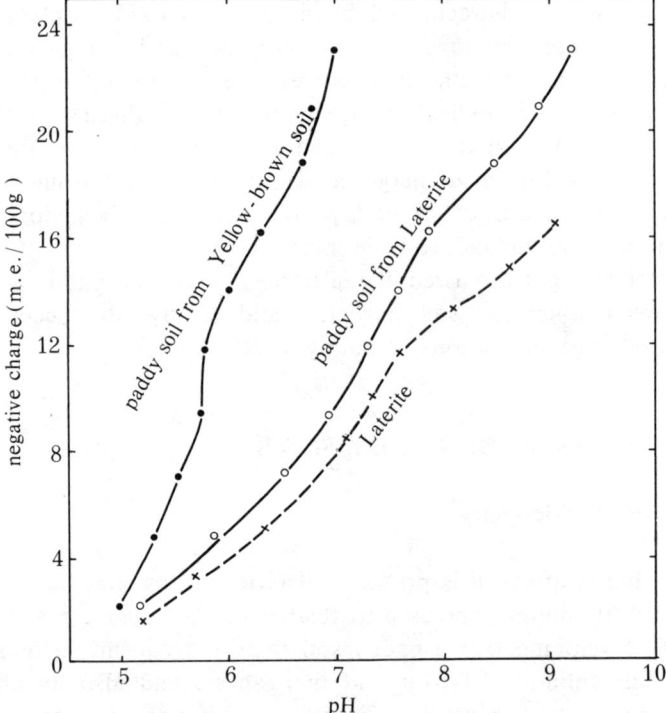

Fig. 6-1 Negative charge of paddy soil and its parent soil

6.1.2 Quantity of electric charge

It has been mentioned that the quantity of electric charge in a soil is remarkably affected by the pH of the medium. In order to compare the quantity of electric charge among different soils on a common basis, it is a usual practice to take the cation exchange capacity at pH 7 as the quantity of negative charge in the soil. This is also adopted in this chapter. At a pH of 7 the positive charge is generally very small in amount or does not exist.

6.1.2.1 Contribution of organic matter to negative charge

Because a part of the organic fraction of the soil is bound to the inorganic fraction to form complexes, it is difficult to accurately determine the quantity of negative charge carried by organic matter individually. Usually it is calculated on the basis of the decrease in quantity of negative charge after the removal of organic matter and the amount of organic matter removed. Obviously this is only an approximate or apparent value.

According to results from over thirty paddy soils, it seems that the quantity of apparent negative charge carried by organic matter is related to the type of the soil. For example, the negative charge calculated from the paddy soils derived from river or lake deposits in Central China averages 126 m.e. per 100 grams of organic matter, whereas that from the paddy soils derived from red soils in South China averages 60 m.e. This seems to imply that the nature of organic matter of soils in different regions is different.

It has been calculated that the contribution of organic matter to the negative charge of the whole soil is 5—42%, with an average of 21%. Generally speaking the contribution in paddy soils is greater than that in their parent soils, due to the increase in organic matter content during the development of paddy soil.

6.1.2.2 Electric charge in different size fractions

The negative charge of paddy soils is mainly concentrated in the clay fraction. However, larger size fractions can also carry a certain amount of negative charge. This is because these fractions can contain some minerals such as kaolinite and illite and also a small amount of organic matter. The contribution of various fractions to the total charge of the soil is dependent on the quantity of negative charge carried by each fraction and the percentage of that fraction. It is seen from Table 6-1 that the contribution of the clay fraction (with a diameter of less than 2 microns) to the negative charge of the soil is higher than 80%, and the remaining 10—15% is contributed by the fractions of 2—10 micron in size.

TABLE 6-1 Contribution of various size fractions to negative charge of paddy soil

Parent material	Locality	Negative charge (m.e. / 100g)				Contribution to total charge (%)			
		<2*	2—10	10—20	20—100	<2	2—10	10—20	20—100
Limestone	Yishan	16.6	6.3	3.9	1.4	81.5	14.2	2.7	1.6
Red clay	Nanning	18.4	2.4	1.3	1.4	82.6	13.3	1.4	2.7
Red clay	Nanning	15.8	3.4	2.9	1.7	82.2	10.6	3.6	3.6
Purplish soil	Qujing	23.2	5.6	5.4	4.5	80.8	12.2	3.2	3.8
Red soil	Zhangzhou	16.0	6.1	2.9	1.4	69.3	25.0	3.5	2.3
Alluvium	Haicheng	26.6	3.2	3.2	4.1	87.1	8.1	2.6	2.2

* unit in μ

6.1.2.3 Negative charge in paddy soils of different regions

The quantity of electric charge carried by paddy soils is affected by a variety of factors, such as the percentage and composition of clay minerals and the amount of organic matter. Because these factors can affect the soil to different degrees even within the same region, it is not surprising that there is no regular difference in the quantity of negative charge carried by the soils in different regions of the country (see Table 6-2). Nevertheless, because of the regularity in zonal distribution of clay minerals, it was found that the quantity of negative charge carried by the clay fraction decreases gradually southward from North China (Table 6-3).

TABLE 6-2 Negative charge of the cultivated layer of paddy soils of various localities[3]

Parent soil	Locality	pH	Clay* (%)	Negative charge (m.e. / 100g)
Alluvial	Yanbian	5.6	—	36.9
Alluvial	Beijing	7.8	—	13.6
Swampy soil	Xinghua	7.6	27.7	19.2
Loess	Nanjing	6.2	19.8	12.5
Lacustrine	Wuxi	5.9	25.0	22.2
Purplish soil	Ganzhou	6.5	36.5	22.3
Red soil	Changsha	5.4	19.0	9.7
Red soil	Nanning	5.8	5.0	8.5
Red soil	Qingyuan	6.1	10.6	10.9
Lateritic soil	Haikang	5.6	28.3	15.5

* $<1\mu$

TABLE 6-3 Negative charge of the clay fraction of paddy soils (cultivated layer) of various localities[2]

Locality	O.M. (%)	Free Fe_2O_3 (%)	K_2O (%)	Negative charge (m.e. / 100g)	Dominant clay minerals
Jilin	3.78	2.83	1.33	52.7	Mt., Ill.
Beijing	3.61	3.98	1.86	40.9	Ill.
Jiangsu	3.04	2.62	2.90	36.6	Ill.
Zhejiang	3.17	2.35	3.55	31.8	Ill.
Jiangxi	2.22	7.65	0.97	25.3	Kl., Ill.
Guangdong	2.14	4.64	1.12	14.1	Kl.
Fujian	1.93	14.0	0.18	17.9	Kl., Sesquioxides

6.1.2.4 Negative charge in relation to fertility status of the soil

The quantity of negative charge carried by a soil may be used as an index of soil fertility. According to determinations for several hundred soil samples, it seems that 13 m.e. of negative charge per 100 g of soil (at pH 7) may be taken as a critical level below which the quantity of electric charge will be of greater importance in affecting the fertility status of the soil. This means that if the negative charge in paddy soil should be smaller in amount, agricultural measures aiming at increasing the quantity of negative charge would be beneficial to the improvement of the fertility status of that soil through increasing its nutrient-holding and -supplying capacity.

6.1.3 Surface charge density

The surface charge density on clays determines the field strength in the vicinity of the clay and is thus intimately related to the electrical double-layer around the clay. Therefore, it can affect many surface properties of the soil. Because of the difference in the composition of clay minerals comprising the micelle and the diversity in the composition and form of organo-mineral coatings covering the micelle, the quantity of electric charge on the clay surfaces and the surface area in various soils will be different. Consequently, the surface charge density will differ not only in different soil clays but also at different sites of the same clay particle. The calculated surface charge density based on quantity of negative charge and total surface area of the soil is only an average value.

It is seen from Table 6-4 that the surface charge density for various paddy soils is around $20\mu C / cm^2$, showing no obvious relationship with the type of the soil. Because the amount of negative charge carried by a soil increases with the rise in pH and the surface area remains practically independent of pH, the surface charge density will increase with the increase in pH. Because it has been found

TABLE 6-4 Surface charge density of the clay fraction of paddy soils[8]

Parent soil	Locality	Negative charge (m.e. / 100g)	Specific surface (m^2 / g)	Charge density (μC / cm^2)
Laterite	Xuwen	12.8	68.5	18.0
Laterite (old)	Xuwen	24.3	110	21.3
Lateritic soil	Bolo	11.2	63.0	17.2
Lateritic soil	Bolo	13.2	65.0	20.0
Swamp soil	Xinghua	43.4	182	23.0
Loess	Jiangsu	44.0	253	16.8

that the removal of iron oxides from the soil resulted in an increase in negative charge density, and there is a distinct loss of iron oxides in the genetic process of paddy soil, there is reason to believe that the negative charge density on the clay surface of a paddy soil would be higher than that of its parent soil.

6.2 INTERACTIONS BETWEEN IONS AND SOIL PARTICLES

6.2.1 Ion adsorption in relation to electric charge of the soil

6.2.1.1 *Adsorption of potassium*

Adsorption of cations is related to the negative charge, but not necessarily to the net negative charge of the soil. It is shown from Fig. 6–2 that the amount of K^+ ions adsorbed by soils increases with the rise in pH. The adsorption also increases with the increase in electrolyte concentration, but the percentage of adsorbed K^+ in the total K^+ present is higher at a lower electrolyte concentration (Fig. 6–3). For example, the percentage of K^+ adsorbed from a 0.001 M solution is higher by over two times than that from a 0.01 M solution. If the two soils are compared it will be found that the amount of K^+ adsorbed by the paddy soil derived from yellow–brown soil is higher than that derived from lateritic soil, due to the larger amount of negative charge carried by the former soil. The quantity of negative charge carried by a soil increases with the rise in pH, hence the amount of adsorbed K^+ also increases, though not proportionally.

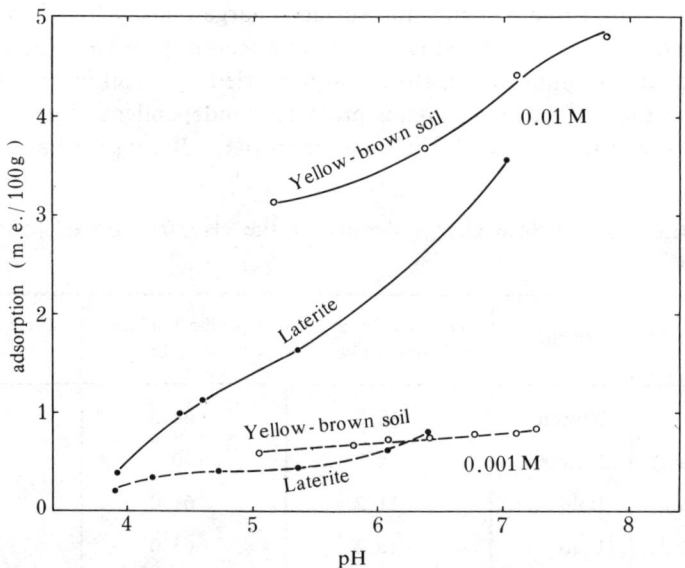

Fig. 6–2 Adsorption of K^+ ions by paddy soils at different K^+ concentrations (soil name in the figures of this chapter denotes parent soil from which the paddy soil is derived)

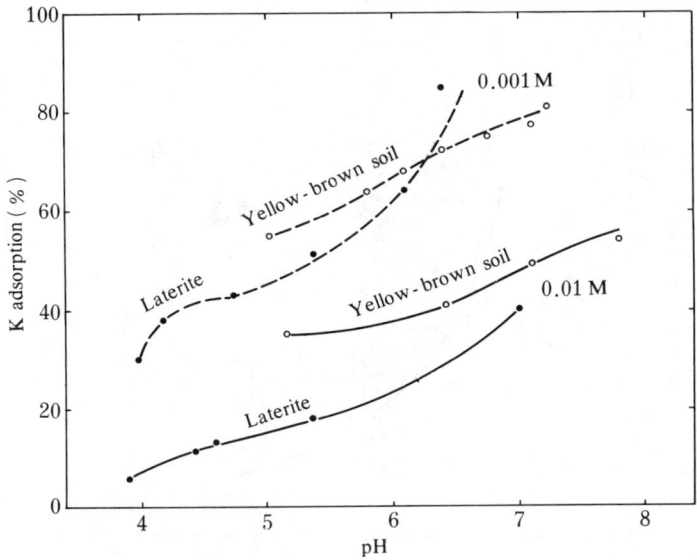

Fig. 6–3 Percentage of adsorption of K⁺ ions by paddy soils at different K⁺ concentrations

6.2.1.2 *Adsorption of chloride*

Adsorption of chloride ions is related to the positive charge, but not necessarily to the net positive charge of the soil. The adsorption decreases with the rise in pH, and eventually changes to negative adsorption. The pH at which there is neither adsorption nor negative adsorption of Cl⁻ may be called the point of zero adsorption. It can be seen from Fig. 6–4 that the point (pH) of zero adsorption for a paddy soil derived from lateritic soil is slightly higher than that for a paddy soil derived from laterite.

6.2.1.3 *Iso-ionic point*

Soils simultaneously carrying negative charge and positive charge can adsorb both cations and anions, being increase in the adsorbed amount of cation with the increase in pH and of anion with the decrease in pH. Some soils can adsorb equivalent amounts of cations and anions at a certain pH, and that pH can be called the iso–ionic point (P_i). In addition to the electric charge of the soil, the iso–ionic point is dependent on the kind and concentration of the cation and anion concerned[10]. It can be seen from Fig. 6–4 that the iso–ionic points with respect to K⁺ and Cl⁻ for a paddy soil derived from laterite, and a paddy soil derived from lateritic soil, are 4.4 and 5.1 respectively.

6.2.2 **Bonding strength of cations with the soil**

When the adsorption–dissociation equilibrium between cations adsorbed

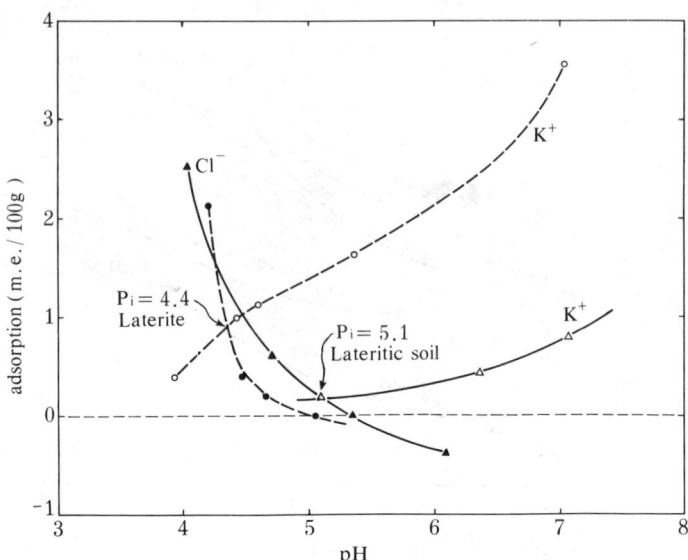

Fig. 6-4 Iso-ionic point (P_i) of two paddy soils with respect to K^+ and Cl^- adsorption

and soil particles carrying negative charge is reached, the magnitude of interacting force between them may be expressed by means of free bonding energy (ΔF), called bonding energy for short. The bonding energy is calculated from the equation

$$\Delta F = RT \ln \frac{1}{f} = RT \ln \frac{c}{a}$$

where f is the fraction active, c, the concentration and a, the activity of the cation in the system. This bonding energy is in reality the electrostatic energy between soil particles and cations, its magnitude being determined mainly by the electric charge on soil particles and the kind and concentration of the cation, and being also related to the kind of complementary cations and the pH of the medium.

Fig. 6-5 shows the bonding energy of K^+ ions in relation to the pH for two paddy soils. For a paddy soil derived from yellow-brown soil, the bonding energy $(\Delta F)_K$ increases with the increase in pH at a constant concentration of K^+. The bonding energy is higher at a lower K^+ concentration (0.001 M) than at a higher concentration (0.01 M) by a factor of bout two. The relationship between bonding energy and the pH for a paddy soil derived from laterite is similar to that derived from yellow-brown soil, except that the absolute value of $(\Delta F)_K$ of the former is lower.

The bonding energy $(\Delta F)_K$ and fraction active f_K at different K^+ saturation percentages can be calculated from a potentiometric titration curve of a H,Al-saturated soil titrated with KOH. Fig. 6-6 shows the results from two paddy soils. For the paddy soil derived from yellow-brown soil the f_K value is quite small.

6.2 Interactions between ions and soil particles

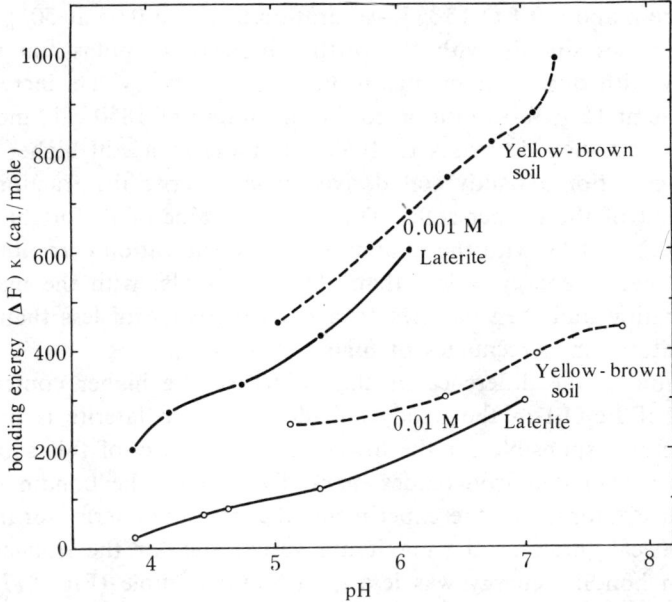

Fig. 6-5 Bonding energy of K⁺ with paddy soils at different pH and K⁺ concentrations

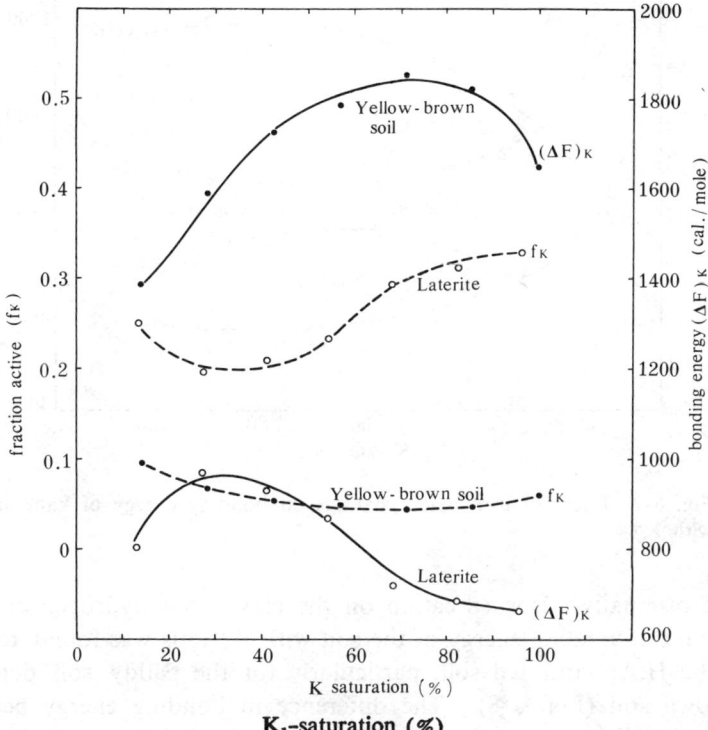

Fig. 6-6 Bonding energy of paddy soils with K⁺ at different K-saturation percentages

It decreases from about 0.1 at 15% K-saturation to about 0.05 at 50% K-saturation, and increases slightly with the further increase in potassium saturation. In conformity with this is the change in bonding energy, which increases from 1400 cal / mole at 15% K-saturation to the maximum of 1850 cal / mole at 70% K-saturation, and then decreases to 1650 cal / mole at a still higher K-saturation percentage. For a paddy soil derived from laterite the fraction active is larger than that of the former soil. The absolute value of f_K for this soil is in the range of 0.2 to 0.33, with the minimum at a K-saturation percentage of 30—40%. The bonding energy is less than 1000 cal / mole, with the maximum at 30% K-saturation and then declines to a constant value of less than 700 cal / mole at K-saturation percentages of higher than 60%.

In addition to the difference in clay minerals, the higher content of iron oxides (11% of Fe_2O_3) in the paddy soil derived from laterite is probably an important factor responsible for the lower bonding energy of this soil. An experiment has shown that iron oxides markedly affected the bonding energy of kaolinite with K^+ ions. In the experiment, the bonding energy for the original clay was 1200 cal / mole as the maximum value, and for the iron-coated clay the maximum bonding energy was less than 800 cal / mole (Fig. 6–7).

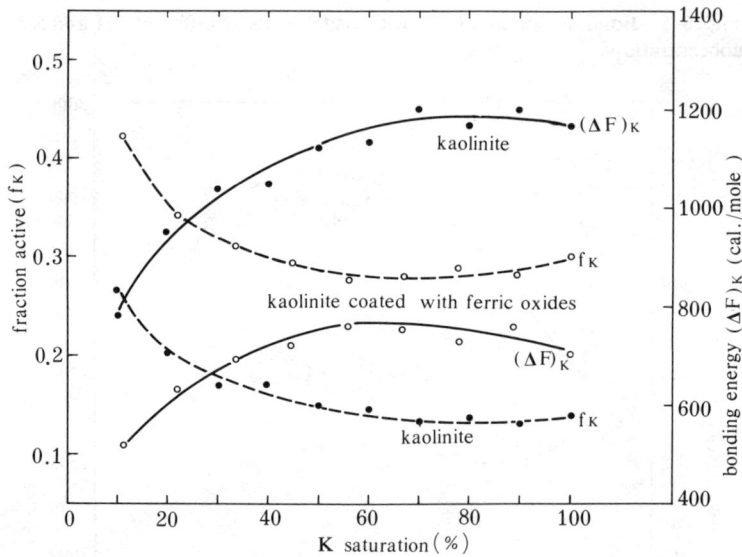

Fig. 6–7 Effect of iron oxide coatings on bonding energy of kaolinite with K^+

If the originally adsorbed cation on the clay is not hydrogen or aluminum but calcium, the bonding energy of the soil with K^+ ions was found to be smaller than the H,Al-saturated soil, particularly for the paddy soil derived from yellow-brown soil (Fig. 6–8). The difference in bonding energy between two Ca-saturated soils was smaller than that between H,Al-saturated soils.

When a Ca-saturated soil is titrated with KOH, the system is actually a Ca-K-saturated system. It was found that the activity ratio a_K/a_{Ca} increases

6.2 Interactions between ions and soil particles

with the increase in degree of K-saturation (Fig. 6-9). The form and slope of the curve were related to the electric charge of the soil and experimental conditions.

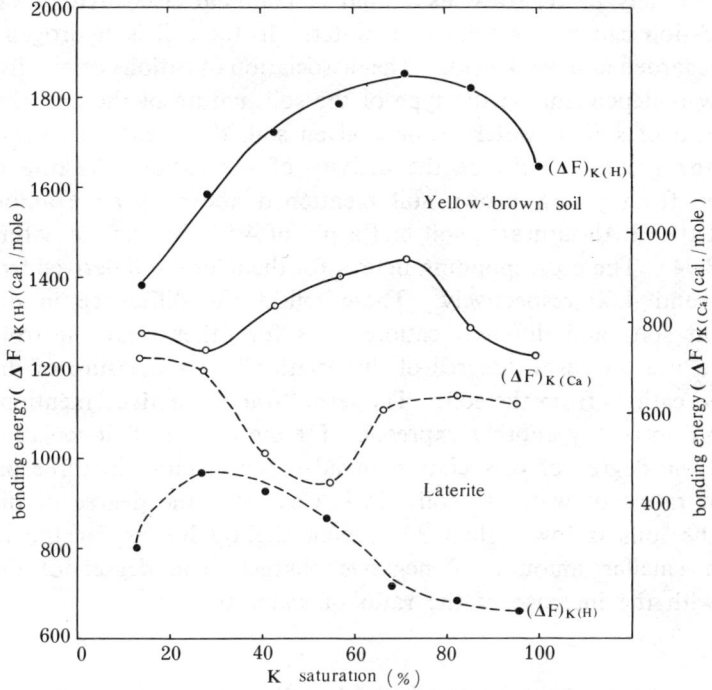

Fig. 6-8 Effect of complementary cations on bonding energy of K^+

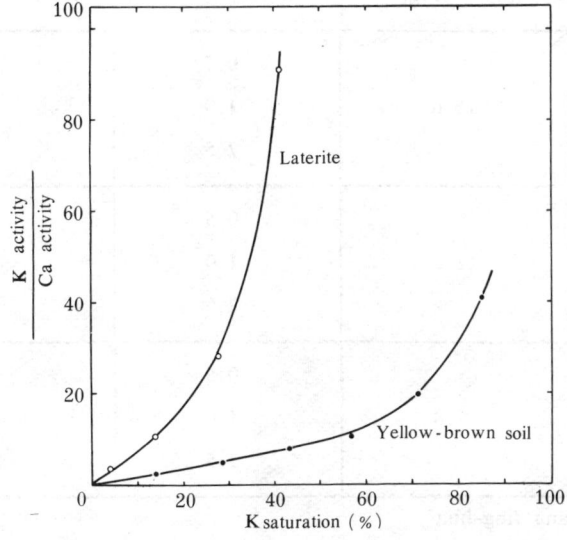

Fig. 6-9 Relationship between K/Ca activity ratio and K-saturation percentage

6.2.3 Dissociation of adsorbed cations

Soil clay may be regarded as a kind of colloidal electrolyte in which the adsorbed cation can be dissociated in water. If the soil is hydrogen-saturated it may be regarded as a weak acid. The dissociation of cations originally adsorbed on the clay is dependent on the type of the soil, nature of the adsorbed cation and the ratio of soil to water. For a given soil, the higher the degree of dissociation the greater would be the activity of the cation. Taking the paddy soil derived from yellow-brown soil mentioned above as an example, it was found that the H,Al-saturated soil had a pH of 4.59 and the Ca-saturated soil, a pCa of 3.74. The corresponding figures for the paddy soil derived from laterite were 4.19 and 4.00 respectively. These reflect the difference in dissociation for different soils and different cations. As for other weak electrolytes, it is common to use the term "degree of dissociation" as a measure of dissociation of adsorbed cations from the soil. The term "fraction active" mentioned in the last section is in reality another expression for the degree of dissociation. Table 6-5 shows the degree of dissociation of adsorbed calcium in three paddy soils at different ratios of water to soil. It is clear that the degree of dissociation for all of the soils is lower than 3%, being slightly higher for the paddy soil carrying a smaller amount of negative charge. The degree of dissociation increases with the increase of the ratio of water to soil.

TABLE 6-5 Dissociation of adsorbed Ca from paddy soils*

Parent soil	C.E.C (m.e. / 100g)	Water to soil ratio	Degree of dissociation (%)
Yellow-brown soil	15.6	0.5	0.74
		1.0	1.46
		2.5	1.74
Red soil	11.7	0.5	1.00
		1.0	1.36
		2.5	2.52
Laterite	7.8	0.5	1.17
		1.0	2.22
		2.5	2.67

* Data from Wang Jing-hua

The dissociation of adsorbed cations can also manifest itself in the electrical conductivity of the system. From conductometric titration curves of two paddy soils (Fig. 6–10) it can be seen that following the increase in percentage of K-saturation the degree of dissociation increases, and, as a result, the specific conductance increases with the increase in the activity of K^+ ions. However, the form of the conductivity curve is slightly different from that of the K^+-activity curve. This is because the electrical conductivity of a soil is also affected by other factors of the system such as pH. It should be noticed that the changes in K^+ activity and electrical conductivity for the paddy soil derived from yellow–brown soil are less than those for the paddy soil derived from laterite. This difference in electrochemical behavior between the two types of soil is presumably related to the difference in content of iron oxides which is only 1.89 % for the former soil and more than 10% for the latter soil. An auxiliary experiment demonstrating the influence of iron oxides on the dissociation of adsorbed K^+ ions showed that the coating of kaolinite with 2—3% of iron oxides caused the K^+ activity and specific conductance of the clay to increase markedly (Fig. 6–11).

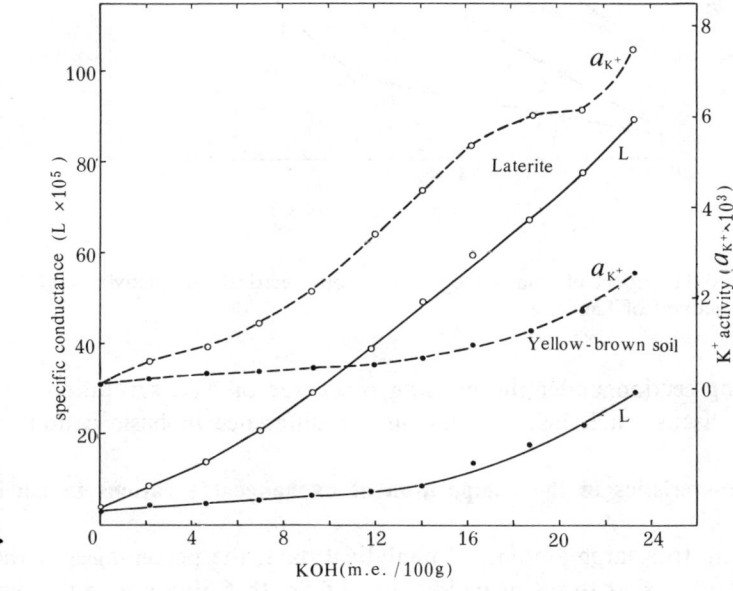

Fig. 6–10 Conductometric and K^+ activity titration curves of paddy soils

6.3 COMPOSITION OF EXCHANGEABLE CATIONS IN PADDY SOILS

Exchangeable cations are the most active components of the soil and are very liable to be affected by natural factors and man–made measures. It is for this reason that the composition of exchangeable cations in various types of paddy soil differs greatly. This difference is mainly reflected in the relative proportions of bases and hydrogen and aluminum ions, which will be discussed

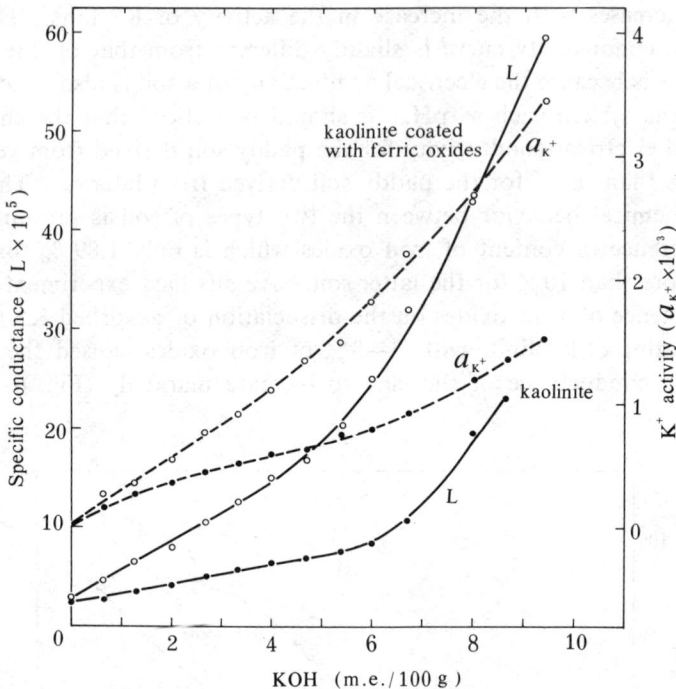

Fig. 6-11 Effect of iron oxide coatings on electrical conductivity and K$^+$ activity of kaolinite

in the ensuing section under the heading of degree of base saturation. In this section, the discussion focuses chiefly on the difference in basic cations.

6.3.1 Characteristics in the composition of exchangeable cations in paddy soil

According to a large amount of analytical data, the percentages of the four most important basic cations in paddy soils of South China are as follows: calcium 50—80%, magnesium 10—30%, potassium 3—6%, sodium 3—10%. The paddy soil is characterized by high percentages of exchangeable potassium and sodium as compared with the upland soil, especially in the cultivated layer. Unlike upland soils, the percentage of sodium in paddy soil is higher than that of potassium. The high percentages of sodium and potassium in paddy soils are apparently related to the frequent application of farm manures.

In addition to the four cations mentioned above, there are usually several milligrams of exchangeable ammonium per 100 grams of soil during the submergence season, corresponding to 1—5% of the total basic cations. It should be remembered that in upland soils there is hardly any exchangeable ammonium.

Besides the basic cations, there is also a certain amount of exchangeable ferrous and manganous ions in submerged paddy soils.

6.3 Composition of exchangeable cations in paddy soils

It has been shown in Chapter 4 that a part of the ferrous iron found under reduced conditions can exchange with originally adsorbed cations on soil particles. Table 6-6 demonstrates the exchange reaction between ferrous ions of the solution and calcium ions adsorbed on the soil. Because the equivalent ratio of adsorbed ferrous iron to replaced calcium was found to be close to 1, it must be concluded that the behavior of ferrous iron with respect to the exchange reaction in paddy soils obeys the general law of equivalent exchange. In submerged soils the exchangeable iron is in dynamic equilibrium with other forms of ferrous iron, and its absolute amount is usually under 0.5 m.e. / 100g but in some cases as high as 2—3 m.e. per 100 g soil (cf. Chapter 4).

TABLE 6-6 Exchange between Fe^{2+} and Ca^{2+} ions in paddy soils[9]

Parent soil	C.E.C. (m.e. / 100g)	Fe^{2+} added (symmetry value)	Fe^{2+} adsorbed (m.e. / 100g)	Ca^{2+} displaced (m.e. / 100g)	$\dfrac{Fe^{2+}\text{adsorbed}}{Ca^{2+}\text{displaced}}$
Yellow soil	8.82	1.0	3.10	3.52	0.88
		2.0	3.44	3.67	0.94
		4.0	5.17	5.43	0.95
Red clay	9.32	1.0	3.10	2.85	1.09
		2.0	3.47	3.23	1.07
		4.0	4.83	4.61	1.05
Red soil	7.36	2.5	4.14	4.39	0.94
		5.0	5.17	5.17	1.00
		10.0	6.89	6.38	1.05

In acid paddy soils there are usually 2—10 mg of exchangeable manganese per 100 g soil even under unsubmerged conditions[5]. If the soil is assumed to contain 5 mg / 100g of exchangeable manganese and the cation exchange capacity of the soil is 8 m.e. / 100g, it can be calculated that the exchangeable manganese will be 0.18 m.e. / 100g, corresponding to 2.3% of the cation exchange capacity. This proportion is similar to that of exchangeable potassium or sodium in upland soils. Under submerged conditions the content of exchangeable manganese would be much higher.

6.3.2 Composition of exchangeable bases in different types of paddy soil

It has already been mentioned that the difference in the composition of exchangeable cations in paddy soils of South China is mainly reflected in the relative proportions of bases and hydrogen and aluminum ions. On the other hand, there is also evidence showing that the composition of exchangeable bases varies in paddy soils derived from different parent materials[1]. In low-fertility paddy soils derived from laterite the proportion of calcium in exchangeable bases is about 50%, with the remaining part contributed by about 30% magnesium, 4—5% potassium and 10—20% sodium. In paddy soils derived from Quaternary red clay the percentages of calcium, magnesium, potassium and sodium are about 60—75%, 15—25%, 3—8% and 4—10% respectively. In paddy soils derived from granite materials the figures for calcium, magnesium and potassium and sodium are about 70—80%, 10—20% and 10% respectively. One peculiarity for paddy soils derived from phyllites is that magnesium accounts for about 45% of the total exchangeable bases, nearly equivalent to calcium.

The proportion of exchangeable potassium in the cultivated layer is generally higher than that in lower horizons.

According to the statistics based on analytical data for over seventy soil samples of South China[1], there is a positive correlation between the ratio of divalent bases to monovalent bases and the base-saturation percentage. However, the ratio is different for paddy soils derived from different parent materials, being lower in those derived from granite and phyllite materials and higher in those derived from laterite and Quaternary red clay.

In neutral paddy soils the calcium-saturation percentage is in the range of 65—80%, with exchangeable magnesium, potassium and sodium in ranges of 13—32%, 0.5—2.6% and 2.0—6.5% respectively[11]. These figures show a higher percentage of calcium or magnesium and a lower percentage of potassium or sodium as compared with acid paddy soils.

6.4 BASE-SATURATION PERCENTAGE IN PADDY SOILS

6.4.1 Base-saturation percentage in relation to pH

On the surface of clay particles of unsaturated soil there are adsorbed base cations as well as adsorbed hydrogen and aluminum ions. A difference in the proportion of these two kinds of cations will result in a difference in base-saturation percentage, and will be manifested in a difference in pH. Fig. 6–12 shows the relationship between base (Ca, Mg, K and Na)-saturation percentage and pH for paddy soils of South China derived from a variety of parent soils. Generally speaking, the effect of base-saturation percentage on pH of the soil is larger at very low or high base-saturation percentages, while it is smaller in the middle range of base-saturation, a fact which is conformable to the general rule for weak acids neutralized with alkalies. The pH at which the soil is completely saturated with bases is around 7, and about 5.3 at a base-saturation percentage of 50% (half-neutralization). It will also be noticed that many points are scattered from the pH–base-saturation percentage curve to varied degrees. This is due to the differences in the composition of exchangeable bases and the holding power of clay minerals for cations in different soils. From Fig. 6–13 it can be seen that the pH of a paddy soil derived from yellow-brown soil is much lower than that derived from laterite at the same K-saturation percentage.

6.4 Base-saturation percentage in paddy soils

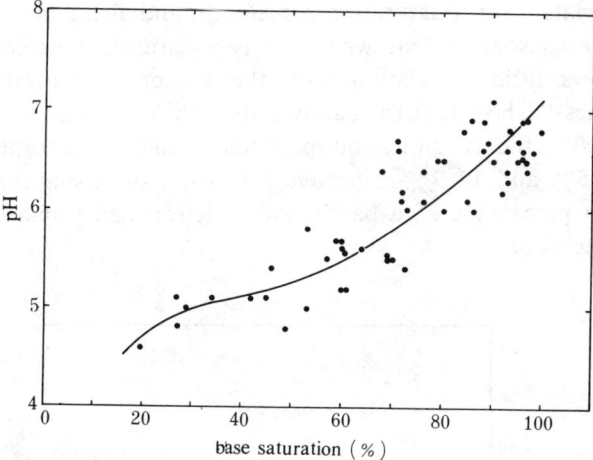

Fig. 6-12 Relationship between pH and base-saturation percentage of paddy soils [1, 6]

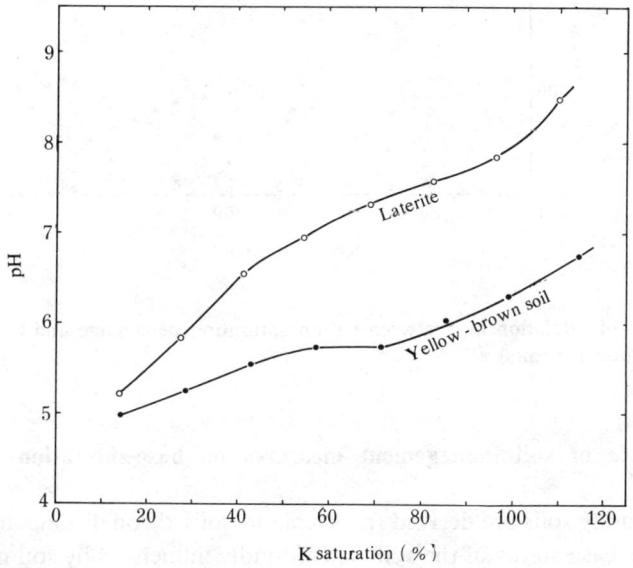

Fig. 6-13 Relationship between pH and K-saturation percentage of paddy soils

6.4.2 Base-saturation percentage in relation to individual cation-saturation percentage

It is interesting to compare the relationship of each cation-saturation percentage of calcium, magnesium, potassium or sodium to the total base-saturation percentage of a soil. Fig. 6-14 shows the results for some representative paddy soils of South China. It can be seen that the base-saturation percentage is clearly related to Ca-saturation percentage, and also related to Mg-saturation percentage to some extent when the base-saturation percentage is below 50%, but shows little correlation with the K- or Na-saturation percentage. Because statistics[1] show that for paddy soils with a base-saturation percentage of 20% to 100% the K-saturation percentage and Na-saturation percentage are only 1—6.5% and 1—9% respectively, it is not surprising that generally the base-saturation percentage of a paddy soil is determined primarily by the Ca-saturation percentage.

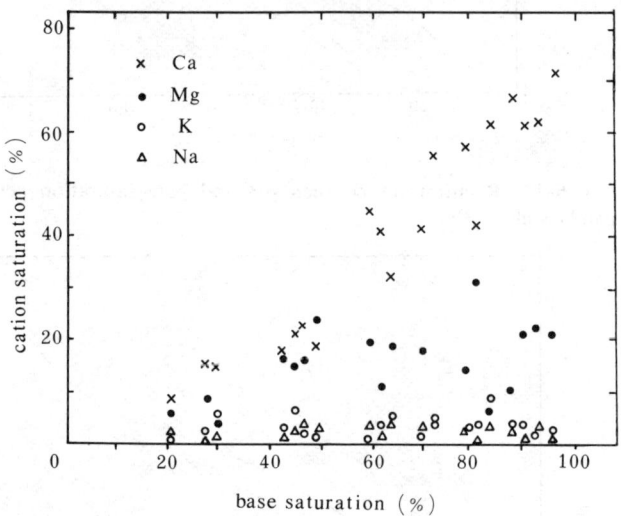

Fig. 6-14 Relationship between cation-saturation percentage and base-saturation percentage[1]

6.4.3 Influence of soil management measures on base-saturation percentage

Because paddy soils are derived from upland soils through long-term cultivation of rice, the base status of the soil is profoundly influenced by soil management measures. For instance, it has been observed that the base-saturation percentage of paddy soils derived from red soils is very low (less than 30%) in exchangeable bases and varies with the age of the soil. For a young paddy soil under cultivation for a few years, the base-saturation percentage is slightly higher only in the cultivated layer, leaving the percentage in horizons below 40—50 cm not different from the parent soil. For a moderately developed paddy soil, the base-saturation percentage in the middle part of the profile is high. For well-developed paddy soils, the base-saturation percentage is high throughout the profile, due apparently to the illuviation of bases from the cultivated layer to which they are added through manuring and irrigation.

It can also be frequently observed that for the same type of paddy soil the base-saturation percentage is closely related to the fertility status of the soil, that is to say, the higher the fertility level, the higher the base-saturation percentage. Fig. 6-15 shows the base-saturation percentage of four paddy profiles derived from Quaternary red clay. In Changsha and Yiwu districts, the cultivation is intensive and the amount of manures applied annually is high, hence the base-saturation percentage in the two profiles is high. On the contrary, in other two profiles with low soil fertility the base-saturation percentage is low.

Neutral paddy soils are generally base-saturated. However, if the field is fertilized with physiologically acid fertilizers continually, the cultivated layer may be base-unsaturated and become an acid soil.

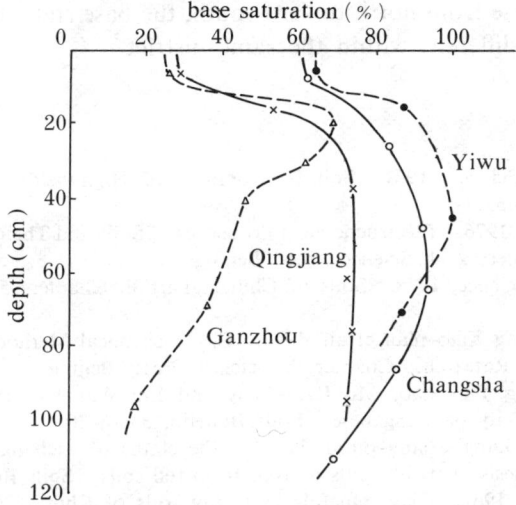

Fig. 6-15 Base-saturation percentage of paddy soils in different districts derived from the same parent soil[6]

6.4.4 Base-saturation percentage of principal types of paddy soil

According to the ratio of exchangeable bases to negative charge carried by the soil, paddy soils may be distinguished as saturated, unsaturated and super-saturated. In saturated soils negative charges are almost completely neutralized by bases such as calcium, magnesium, potassium and sodium. In unsaturated soils, in addition to exchangeable bases, the soil still contains hydrogen and aluminum ions, and thus the base-saturation percentage is lower than 100%. In super-saturated soils the soil contains surplus bases which are present in the form of water-soluble salts (such as NaCl), or difficultly soluble salts (such as $CaCO_3$). The base status of paddy soils is determined primarily by the degree of leaching during the genesis of the paddy soil as well as its parent soil, and is also influenced by man-made measures. Generally speaking, paddy soils which are weakly leached are super-saturated, those which are moderately leached, are base-saturated, and those which are strongly leached, are unsaturated. From the viewpoint of zonal distribution, most of the paddy soils in North China are supersaturated, the majority of paddy soils in Central China are nearly base-saturated, and a large part of the paddy soils in South China are unsaturated. However, there are many exceptions. For instance, in the Sunghua River plain of Northeast China, most of the soils are base-saturated or even unsaturated, whereas in the lower reaches of main rivers and in coastal regions of South China, there are large areas of paddy soil with a nearly saturated or supersaturated base status. Some young paddy soils derived from calcareous purplish soils in South China are also base-saturated or supersaturated. In some districts of South China there are even paddy soils containing calcium carbonate as a result of overliming. Therefore, in China although there is a general tendency for the base-saturation percentage to decrease from north toward south, the base status of various paddy soils may be quite different within the same district.

REFERENCES

(1) Institute of Soil Science, 1961. Soil Environment of High-yield Rice. Chapters 5, 9. Science Press, Beijing.
(2) Yu Tian-ren et al., 1976. Electrochemical Properties of Soils and Their Research Methods. (revised ed.) Chapters 2, 4. Science Press, Beijing.
(3) Institute of Soil Science, 1978. Soils of China. Part B, Chapters 5, 8. Science Press, Beijing.
(4) Yu Tian-ren, Zhang Xiao-nian et al., 1980. Electrochemical Methods and Their Applications in Soil Research. Chapter 9. Science Press, Beijing.
(5) Yu Tian-ren, Ling Yun-xiao, Mu Ren-sheng and Liu Wan-lan, 1958. Effect of soil acidity on the activity of manganese. Soils Bulletin, 33:16–30.
(6) Yu Tian-ren and Ding Chang-pu, 1958. On the status of exchangeable bases and its relation to the genesis of paddy soils derived from red soils. Soils Bulletin, 33: 31–43.
(7) Zhang Xiao-nian, 1961. Clay minerals in paddy soils of China. Acta Pedologica, **9**: 81–102.
(8) Zhang Xiao-nian and Jiang Neng-hui, 1964. Studies on electrochemical properties of soils. III. Electric charges of the clay fraction of red soils. Acta Pedologica, **12**: 120–131.
(9) Bao Xue-ming, Liu Zhi-guang, Wu Jun and Yu Tian-ren, 1964. Studies on oxidation-reduction processes in paddy soils. VII. Forms of ferrous iron. Acta Pedologica, **12**: 297–306.
(10) Zhang Xiao-nian, Jiang Neng-hui, Shao Zung-chen, Pan Shu-zheng and Zhang Wan-gen, 1979. Studies on electrochemical properties of soils. VI. Adsorption of ions by red soils in relation to the electric charge of the soil. Acta Pedologica, **16**: 145–156.
(11) Chen Jia-fang and Shao Zung-chen, 1979. Exchangeable bases in paddy soils of Suzhou district. Soils, **6**: 215–219.

CHAPTER 7

ACIDITY

Cang Dong-qing, Wang Jing-hua and Zhang Xiao-nian

Soil acidity is a comprehensive reflection of many chemical properties of the soil, and it in turn exerts profound influences on a series of other properties relating to soil fertility. The acidity of paddy soils is characterized by a periodical change induced by alternate irrigation and drainage, and the rate and extent of such changes are far greater than those occurring in upland soils. In this chapter, in addition to general properties of soil acidity as related to paddy soil, this characteristic feature of paddy soil will be discussed.

7.1 INDEXES OF SOIL ACIDITY

In the study of soil there are several indexes for characterizing the acidity status of a soil. In this chapter only pH, lime potential, pK, and exchange acidity and exchange alkalinity will be discussed. These indexes are also closely interrelated.

7.1.1 pH

There are a variety of factors which can affect the pH of a soil. The role of the composition of adsorbed cations has already been mentioned in Chapter 6. Because for the same soil the pH may vary considerably with the variations in water content, kind and concentration of electrolytes, partial pressure of carbon dioxide, etc., and there are always unavoidable measurement errors due to technical difficulties, the pH value as determined by conventional methods is only conditional. In the following, the effects of the ratio of water to soil and electrolyte concentration will be considered.

7.1.1.1 *Effect of the ratio of water to soil*

Because of the effects of water on the distribution of various ions between the solid phase and the liquid phase of the soil and on the solubility of some salts such as calcium carbonate, the pH of a soil will vary with the variation in water content. Fig. 7-1 shows some results for paddy soils derived from different parent soils. The general tendency is that the pH of a soil increases with the increase in water content. The increase of pH for a calcareous paddy soil is about 1 unit, and for a paddy soil derived from yellow soil is about 0.2 unit. The larger increase in pH for a calcareous paddy soil is due to the hydrolysis of alkaline substances, and the pH increase for the acid paddy soil is the result of dilution effect. For each soil the pH change may be considered in three ranges of water content. Below water-saturation the pH increases sharply and almost linearly with the increase in water content. At the vicinity of water-saturation the curve is flattened, and then the effect of water content lessens until the ratio of water to soil is up to 5, above which the pH is practically independent of the variation in water content.

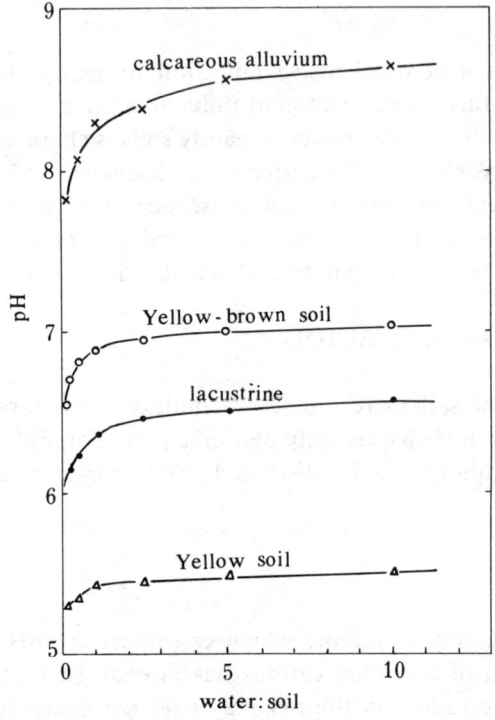

Fig. 7-1 Effect of water to soil ratio on the pH of paddy soils derived from different parent soils[10]

7.1.1.2 *Effect of concentration of neutral salts*

Fig. 7-2 shows the change in pH with the variation in concentration of neutral salt for three soils. It is seen that the pH decreases with the increase in NaCl concentration, especially at high concentrations. The effect of neutral salt on pH of the soil is due chiefly to the replacement of adsorbed hydrogen and aluminum ions from the surface of clay into solution by cations of the salts. For soils carrying positive charge the anion of the salt can exchange with such anions as OH^- adsorbed by the clay. It is thus clear that the direction and extent of pH change in the presence of neutral salts will be different for different soils and different salts. This question will be discussed in a later section under the heading of exchange acidity and exchange alkalinity.

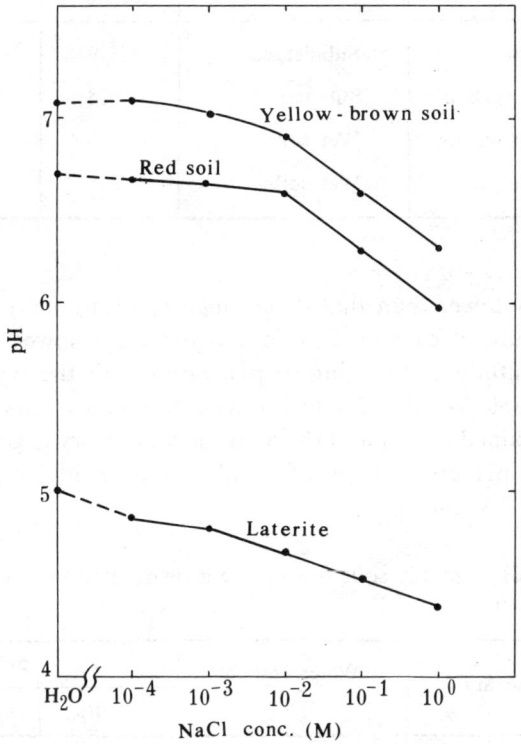

Fig. 7-2 Effect of concentration of neutral salt on the pH of paddy soils

7.1.1.3 *Comparison between soil measurements in situ and sampling*

It is a conventional practice to determine soil pH in the laboratory with soil samples. However, in order to understand the relationship between plant growth and soil conditions it is more desirable to determine the pH of the soil in situ. For, only the pH value in situ is representative of the real pH of the soil under its natural condition and the actual environment of plant roots, which are of more concern to us. Since for paddy soils there is usually a rotation of rice and upland–crops, and in the rice–growing period the water content of the soil may vary greatly, the pH of a paddy soil fluctuates considerably. It can be seen from Table 7-1 that for two submerged paddy soils derived from red soil the pH determined directly in situ is one unit higher than those determined in the laboratory by the conventional method with a water to soil ratio of 1:1. The cause of the increase in pH during submergence is generally assumed to be the consumption of H^+ ions during reduction processes.

During the season when upland crops are grown, the pH of paddy soil deter-

TABLE 7-1 pH of paddy soils during the growing period of rice[10]

Parent soil	Locality	Water regime	pH	
			In situ	Laboratory*
Red soil	Zixi	Submerged	5.9	4.9
Red soil	Hangzhou	Submerged	6.4	5.4
Lateritic soil	Guangzhou	Wet soil	5.3	5.4
Yellow soil	Gaoyao	Wet soil	5.5	5.6

* Water : Soil = 1:1

mined in situ is usually lower than that determined conventionally. This is due to higher partial pressure of carbon dioxide and especially lower water content of the soil. The magnitude of lowering in pH varies with the type of the soil. It can be seen from Table 7-2 that for acid and neutral paddy soils the difference between the results obtained in situ and those in the laboratory is generally within the range of 0.1 to 0.2 pH unit, whereas for calcareous paddy soils the pH in situ may be lower by 0.8 unit.

TABLE 7-2 pH of paddy soils during the growing period of upland crops (Jiangsu)[10]

Parent soil	Locality	Water content (%)	pH	
			In situ	Laboratory*
Red soil	Yixing	32.4	5.4	5.5
Alluvium	Yixing	33.2	6.3	6.5
Lacustrium	Wujin	23.9	6.3	6.4
Loess	Nanjing	saturated	6.9	7.0
Terra rossa	Yixing	36.4	7.6	7.9
Calcareous alluvium	Nanjing	22.9	7.5	8.3
Calcareous alluvium	Nanjing	Sticky point	7.4	—

* Water : Soil = 1:1

7.1.2 Lime potential

The lime potential of a soil is defined as the negative logarithm of the activity of hydrogen ions minus half that of calcium ions, that is, pH–0.5 pCa. In order to calculate the pH–0.5pCa value, the generally used method was to determine the pH with a glass electrode, analyse the solution for calcium concentration,

7.1 Indexes of soil acidity

and then compute in terms of the activity of Ca. This method is not only tedious but is also inevitably subject to errors caused by the inaccuracy in the calculation of the activity coefficient of a single ion species and by a liquid–junction potential in the determination of pH. Recently a method has been devised in which the lime potential is determined directly by inserting a glass pH electrode together with a calcium-selective membrane electrode into the soil system, and satisfactory results have been obtained[9]. In the following section some relevant questions will be discussed.

7.1.2.1 *Lime potential as an index of soil acidity*

From the mathematical formula for lime potential, pH–0.5pCa, it is clear that lime potential is directly correlated with pH and inversely correlated with pCa. Because in soils the pH is generally inversely correlated with pCa, it may be suggested that it is possible to integrate the two values to characterize the acidity of a soil. From the potentiometric titration curves of an acid soil (Fig. 7-3) it is seen that following the addition of $Ca(OH)_2$ the general pattern of the change in pH–0.5pCa is similar to the change in pH, and is even more distinct.

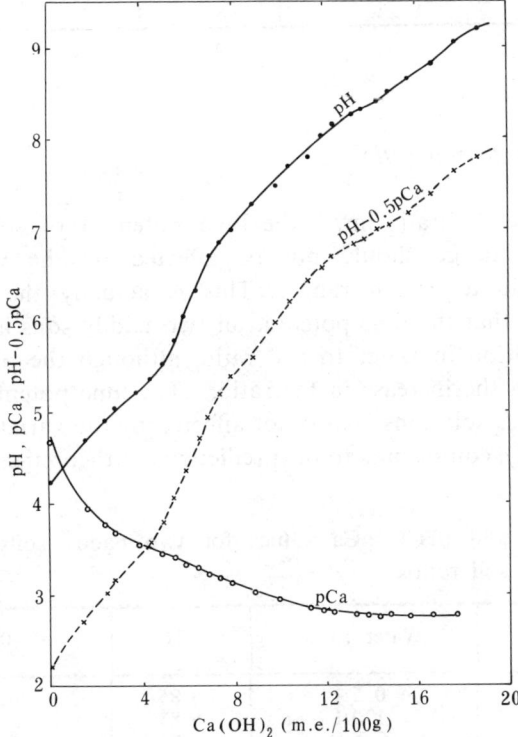

Fig. 7-3 Changes in pH, pCa and pH–0.5pCa in the titration of soil with $Ca(OH)_2$[9]

Table 7-3 shows the pH and lime potential of different types of paddy soil and their parent soil. The difference in lime potential among paddy soils or among their parent soils is greater than the difference in pH. If comparison is made between a paddy soil and its parent soil it will be observed that the numerical values of Δ(pH–0.5pCa), i.e., lime potential of paddy soil minus that of its parent soil, are always larger than those of ΔpH, especially for the laterite and the paddy soil developed on it. Because the difference in lime potential between two soils represents differences in pH and pCa simultaneously, which are generally additive, it is more distinct than the difference in pH or pCa alone.

TABLE 7-3 Comparison between ΔpH and Δ(pH–0.5pCa) for paddy soils and their parent soils*

Soil type	pH			pH–0.5pCa		
	Paddy	Parent	ΔpH	Paddy	Parent	Δ(pH–0.5pCa)
Laterite	5.23	5.12	0.11	3.40	2.29	1.11
Red soil	6.56	5.15	1.41	4.93	3.02	1.91
Yellow–brown soil	6.83	5.71	1.12	5.32	3.91	1.41

* Water : Soil = 1:1

7.1.2.2 Effect of water to soil ratio

According to Schofield's "ratio law", the lime potential of soils carrying predominantly negative charge should not be affected by the variation of water to soil ratio within a certain range. This is actually the case. The data of Table 7-4 shows that the lime potential of two paddy soils is practically independent of the variation in water to soil ratio, although the pH shows a tendency to increase with the increase in the ratio. The lime potential of three paddy soils in dilute $CaCl_2$ solutions is also not affected by the variation in solution to soil ratio (Fig. 7-4), conforming to the predictions of theoretical deduction.

TABLE 7-4 pH and pH–0.5pCa values for two paddy soils at different water to soil ratios

Parent soil	Water : Soil	pH	pH–0.5pCa
Yellow–brown soil	0.5:1 1.0:1 2.5:1	6.85 6.83 7.04	5.34 5.32 5.37
Lateritic soil	0.5:1 1.0:1 2.5:1	5.15 5.23 5.30	3.41 3.40 3.39

7.1 Indexes of soil acidity

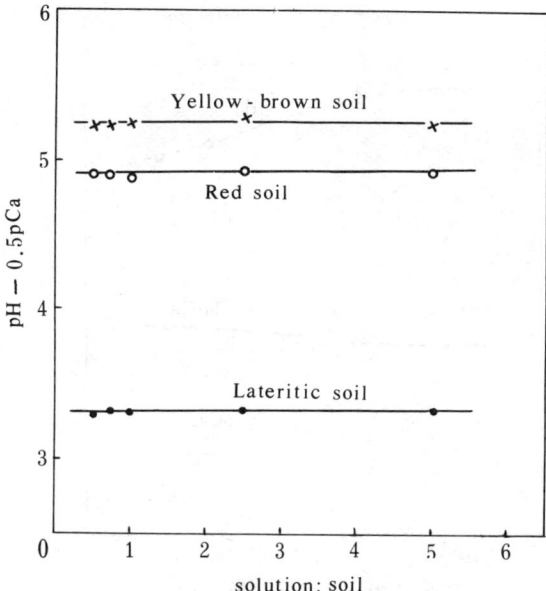

Fig. 7-4 Effect of solution to soil ratio on lime potential of paddy soils (0.001 M $CaCl_2$)[9]

7.1.2.3 *Effect of salt concentration*

The effect of the concentration of calcium salt on lime potential is a combined reflection of the effects on pH and pCa. Generally, the pH and pCa change in the same direction. In Fig. 7-5, for a paddy soil derived from lateritic soil, the lime potential only changes slightly, although the pH and pCa decrease remarkably with the increase in $CaCl_2$ concentration. From the variation of lime potential in $CaCl_2$ solutions of different concentrations shown in Fig. 7-6, it is obvious that for a paddy soil derived from yellow-brown soil and carrying predominantly negative charge the lime potential is not affected by $CaCl_2$ concentration, and for a paddy soil derived from lateritic soil and carrying some positive charge the lime potential increases only at salt concentrations of higher than 0.001 M, whereas the variation is indistinct for a paddy soil derived from red soil.

7.1.3 p*K*

p*K* represents the negative logarithm of the dissociation constant of the acidoid group. According to the principle of chemical equilibrium for weak acids, the pH value at which the acid is half-neutralized (i.e., the ratio of salt/acid is 1) is the p*K* value of the weak acid. Therefore, this value may also be called "half-neutralization pH". The nature of a soil in respect to acid is akin to that

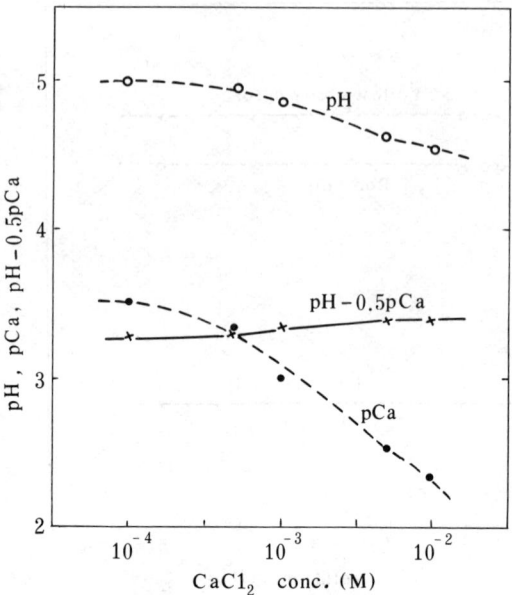

Fig. 7-5 Effect of CaCl$_2$ concentration on pH, pCa and pH–0.5pCa for a paddy soil derived from lateritic soil

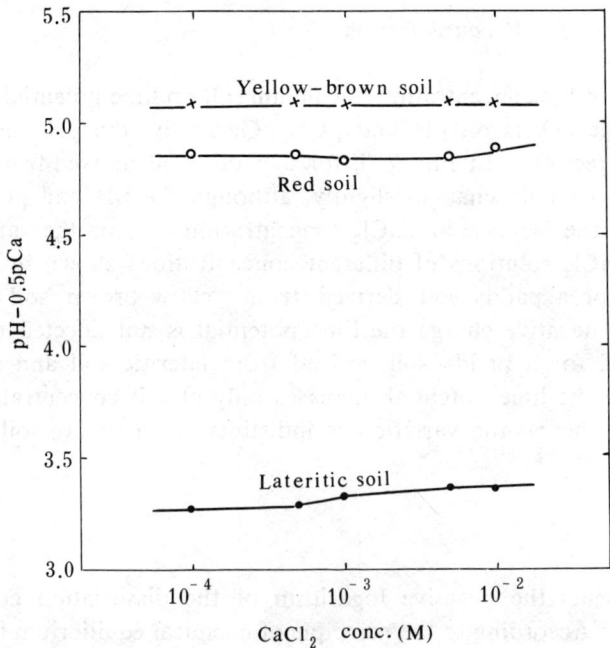

Fig. 7-6 Effect of CaCl$_2$ concentration on pH–0.5pCa for three paddy soils[5]

7.1 Indexes of soil acidity

of a weak acid, and the pH at which the soil is half–neutralized with bases may be regarded as the pK of that soil. Therefore, pK value may be used as an index for expressing the acidity strength of the soil, and its value is smaller if the acidity strength is stronger. Actually a soil clay may be regarded as a polyfunctional acid carrying acidoid groups with different degrees of dissociation, and the ordinarily determined pK value is only an average value.

It can be seen from Table 7–5 that the pK of paddy soils derived from different parent materials in different regions of China differs markedly, showing a zonal regularity of decreasing in numerical value from the south to the north. This difference is assumed to be related to the difference in the composition of clay minerals. It has been known that the pK value of three kinds of soil colloid is of the following order: organic colloid<aluminosilicate<hydrated oxides of iron and aluminum. This means that the ability to furnish hydrogen ions to a soil solution is of the order: organic colloid>aluminosilicate>hydrated oxides of iron and aluminum. It has also been known that the pK value of soil clay carrying permanent negative charge is smaller than that caused by variable negative charge, that is to say, the dissociation of hydrogen and aluminum ions adsorbed by permanent charge is easier than those adsorbed by variable charge. The difference in the mean value of pK shown in Table 7–5 is in reality a reflection of these relations.

TABLE 7–5 Half–neutralization pH (pK value) of paddy soils derived from different parent soils[3]

Parent soil	Locality	No. of samples	pK value (mean)
Lateritic soils	Yunnan, Guangdong	16	5.49
Red and yellow soils	Jiangxi	26	5.46
Purplish soils	Jiangxi, Yunnan	11	5.22
Red soils	Jiangxi, Hunan, Zhejiang	57	5.06
Neutral alluvium	Hubei, Zhejiang	6	4.51

The half–neutralization pH (i.e., pK) of some paddy soils as found from the potentiometric titration curve with NaOH is much higher than that shown in Table 7–5. This is because for these natural soils of Table 7–5, the adsorbed bases are predominantly calcium and magnesium. Thus, the nature of adsorbed basic cations may have some influence on the pK value of a soil.

7.1.4 Exchange acidity and exchange alkalinity

On soil particles carrying negative and positive charge simultaneously there are adsorbed hydrogen and aluminum ions as well as hydroxyl ions, which, once exchanged with cations and anions of a neutral salt, will result in the formation of exchange acidity and exchange alkalinity at the same time. For most of the paddy soils of China, the quantity of exchangeable hydrogen and aluminum ions is much more than that of exchangeable hydroxyl ions; therefore, the pH of a soil in salt solutions is generally lower than that in water. However, for paddy soils derived from laterite and carrying a large amount of positive charge, the quantity of exchange alkalinity may exceed that of exchange acidity, and as a result the pH in a neutral salt solution may be higher than that in water.

The change in pH of a soil in salt solutions is also related to the kind and concentration of the electrolyte. In Fig. 7–7, because the replacing power of SO_4^{2-} for hydroxyl ions is stronger than that of Cl^-, the pH of the soil in a Na_2SO_4 solution is higher than that in a NaCl solution (cf. Fig. 7–2). Conversely, because the replacing power of Ba^{2+} for hydrogen and aluminum ions is stronger than that of Na^+, the pH of the soil in a $BaCl_2$ solution is lower than that in a NaCl solution. When a comparison is made among three paddy soils derived from different parent materials, it will be observed that for the paddy soil derived from laterite the pH in Na_2SO_4 solutions is higher than that in water due to the predominance of adsorbed hydroxyl ions, and for the other two soils the pH is slightly higher only at low concentrations of Na_2SO_4. When soils are

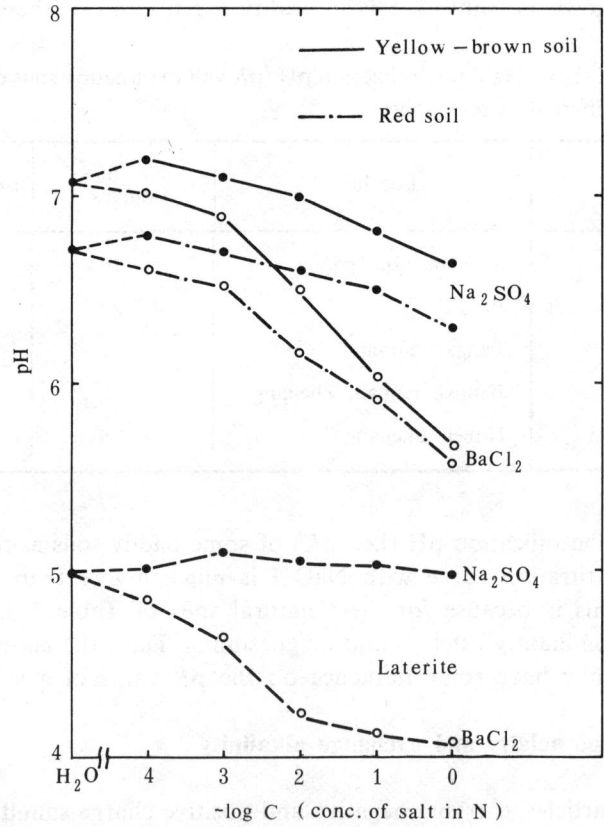

Fig. 7–7 Exchangeable acidity and exchangeable alkalinity of three paddy soils

reacted with $BaCl_2$ solutions, the pH decreases with the increase in salt solution. The decreases in pH for paddy soils derived from yellow–brown soil, red soil and laterite are from 7.1 to 5.7, from 6.7 to 5.6 and from 5.0 to 4.1 respectively.

The exchange acidity and exchange alkalinity of a soil is generally manifested in the form of hydrogen or hydroxyl ions. In reality the exchange acidity of a soil is caused chiefly by exchangeable aluminum[4]. These aluminum ions are exchanged to the solution by cations of neutral salts and then undergo hydrolysis, producing hydrogen ions. It can be seen from the data of Table 7–6 that the contribution of aluminum ions to exchangeable acidity or water–soluble acidity far exceeds that of hydrogen ions.

TABLE 7–6 Composition of exchangeable and water–soluble acidity of paddy soils[3, 8]

Soil	Locality	Depth (cm)	pH	Acid (m.e. / 100g)		
				H	Al	Sum
Clayey (exchange-able acid)	Guizhou	0—18	5.5	—	0.45	0.45
		18—28	4.9	0.18	4.14	4.32
		45—55	4.9	0.15	3.60	3.75
Acid sulfate (water-soluble acid)	Guangdong	0—14	3.1	0.07	0.37	0.44
		14—26	2.8	0.08	1.56	1.64
		26—50	2.3	0.71	10.71	11.42

7.2 CHARACTERISTICS IN THE ACIDITY OF PADDY SOIL

7.2.1 Change in acidity during submergence

The change in acidity during submergence is the most important feature of paddy soil as far as soil acidity is concerned. The direction, extent and rate of this change are closely related to the original pH of the soil and the decomposition pattern of organic matter.

7.2.1.1 *Relation to original pH of the soil*

The magnitude of pH change after submerging differs with the original pH of the soil. It is seen from Fig. 7–8 that for a strongly acid paddy soil the pH increased rapidly, and then attained a steady value about two weeks after submerging. If the original pH of the soil is within the neutral range, the change in pH will be indistinct. For a slightly alkaline soil with an original pH of 8 the pH decreased to a steady value after about two weeks. The patterns of pH change stated above are quite representative of paddy soils. Under field conditions, the acid paddy soil (Table 7-7) and alkaline paddy soil (Table 7-8) tend to change their pH towards neutrality after submerging, namely, from a variation range of 4.6 to 8.0 to a range of 6.6 to 7.5.

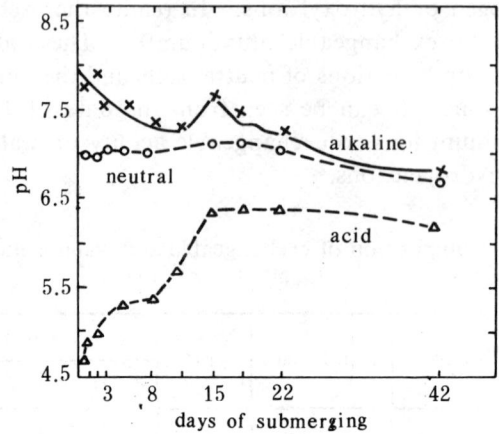

Fig. 7-8 Change in pH after submerging for different types of paddy soil[2]

TABLE 7-7 Change of pH in acid paddy soils after irrigation (Zixi)[1]

Soil	O.M. (%)	pH		
		Before irrigation	4th day	20th day
Wusheng	4.45	5.37	5.68	6.75
Beisha	4.45	6.10	6.55	6.98
Beisha	1.94	6.12	6.25	6.56

TABLE 7-8 Change of pH in an alkaline paddy soil after irrigation (Changshu)[1]

Depth (cm)	pH				
	Before irrigation	10th day	15th day	35th day	43th day
3—10	8.0	7.9	7.5	7.0	6.9
15	—	7.9	7.5	7.2	6.8
25	8.0	8.2	7.6	7.8	7.5

After submerging, the lime potential will also change due to the change in pH and pCa. Fig. 7-9 shows some of the results. For a paddy soil derived from laterite with an original pH of 5 and containing a considerable amount of organic matter, the lime potential increased by about 1.5 units from an original value of about 3 three weeks after submerging. For a paddy soil derived from yellow-brown soil with an original pH of about 7 and containing less organic matter, the increase in lime potential was within 1 unit. For a calcareous paddy soil with an original pH of about 8, the lime potential showed a slightly declining tendency. It is thus clear that as in the case of pH change, the difference in lime potential among various paddy soils tends to become smaller with the increase in duration of submergence, i.e., from a wide range of 3—7 to a narrow range of 4.5—6.5.

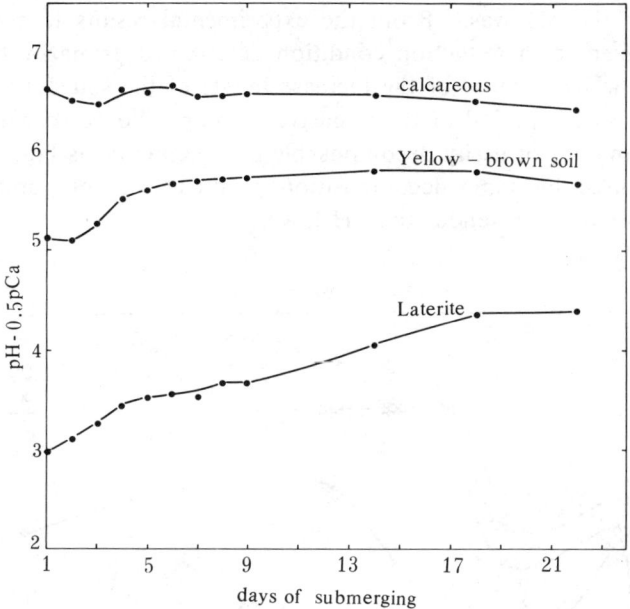

Fig. 7-9   Change in lime potential after submerging for different types of paddy soil

7.2.1.2   *Relation to organic matter*

Organic matter can affect the change of soil acidity during submergence in two ways. The organic reducing substances formed during the decomposition of organic matter may reduce the oxides of iron and manganese, causing the pH to increase due to the consumption of protons in the course of the reduction of oxides. Organic acids and CO_2 produced in the decomposition processes of organic matter may make the pH decrease. The direction of pH change due to submergence is the result of the integrative influences of these two contributing factors. Fig. 7-10 shows the dynamic change of pH of an acid paddy soil during submergence after the addition of green manure. Within the first three days the pH increased abruptly and almost linearly. Within the time interval of three to ten days changes of the pH kept comparatively steady and then increased slowly for treatments with low and medium amounts of green manure. For the treatment with a high amount of green manure there appeared a minimum at about the seventh day and then an increase. When the pH reached a steady value of about 7 after two weeks the effect of dose of green manure became indistinct. For the control treatment, the trend in pH change was similar to that for the treatments of low and medium amounts, with the exception that during the whole submergence period the pH was lower. If lime had been added to the soil the pH-time curve changed gradually from a convex form for the control treatment to a concave form for treatments with an increasing amount of green manure. At a given time, the higher the amount of green manure added, the lower the pH was. From the experimental results it may be said that the development of a reduction condition determined primarily by organic matter status is the basic cause of the increase in pH of the soil during submergence. But at a certain period of the intensive decomposition of organic matter, especially if the amount of easily decomposable organic matter is high, the acidification effect caused by some decomposition products can temporarily retard the rise of soil pH or even render the pH lower.

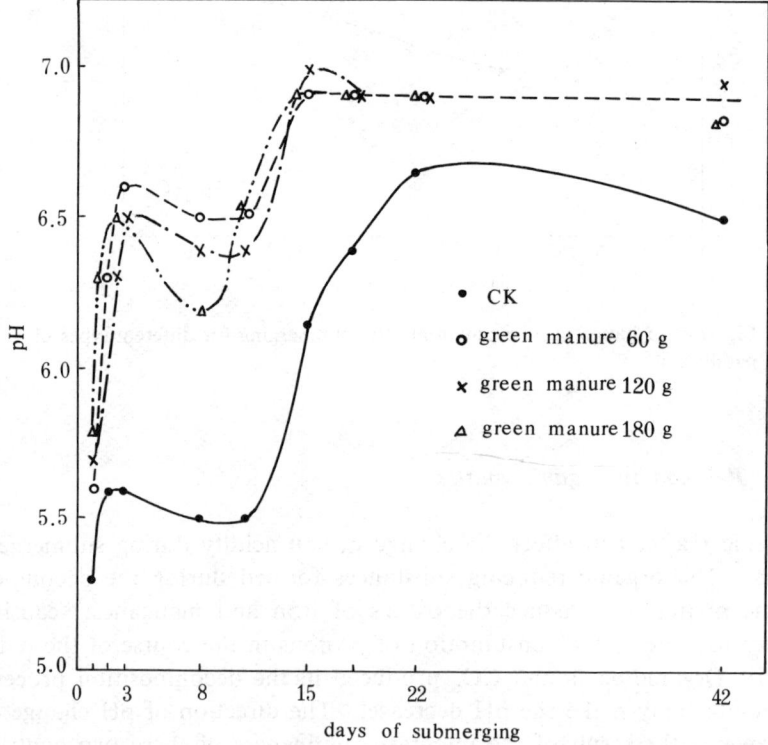

Fig. 7-10 Influence of organic matter on change in pH of acid paddy soil after submerging

7.2.1.3 Effect of liming

Lime can neutralize the active acidity and exchangeable acidity of the soil. Besides, it can also affect soil pH indirectly through promoting the decomposition of organic matter. From the dynamic change in pH of a paddy soil derived from red soil (Fig. 7–11) it can be seen that in the liming treatments (curves 3 and 4) the pH rose rapidly at the second day of submergence due to the neutralization of active acidity. The subsequent decrease in pH to a minimum at the eighth day or so was due to the production of organic acids and CO_2 during the decomposition of organic matter. Still later, the development of the reduction condition caused the pH to rise again slowly to a steady value. In the control treatments without the addition of lime (curves 1 and 2) the rise in pH proceeded more slowly, and the pH was lower than the liming treatments in both the manuring and non–manuring treatments.

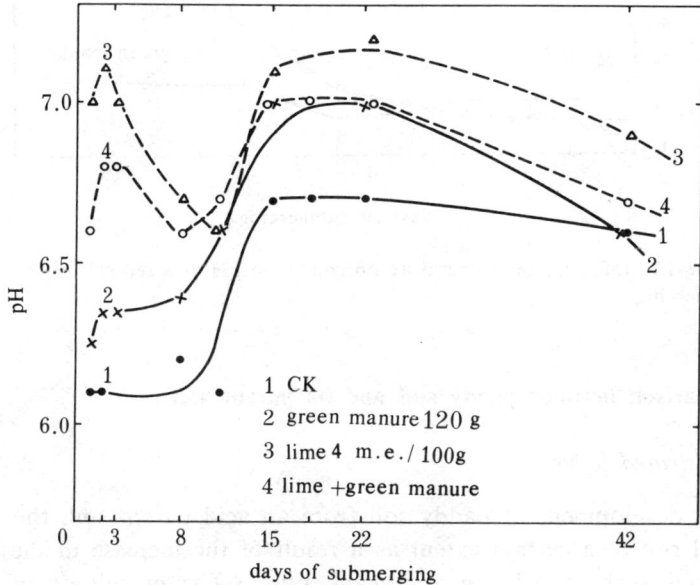

Fig. 7–11 Influence of lime on change in pH of paddy soil (derived from red soil) after submerging

7.2.1.4 Influence of temperature

Temperature influences the change in soil pH through affecting the neutralization rate and the decomposition rate of organic matter. Fig. 7–12 shows that the change in pH is indistinct if the soil is incubated at 15°C, and is remarkable if the soil is incubated at 28°C. Because the magnitude of pH change is greater in the liming treatment than in the unlimed treatment, it may be assumed that the stimulating effect of lime on the decomposition of organic matter can be manifested remarkably only under conditions where the temperature is suitable for the decomposition of organic matter.

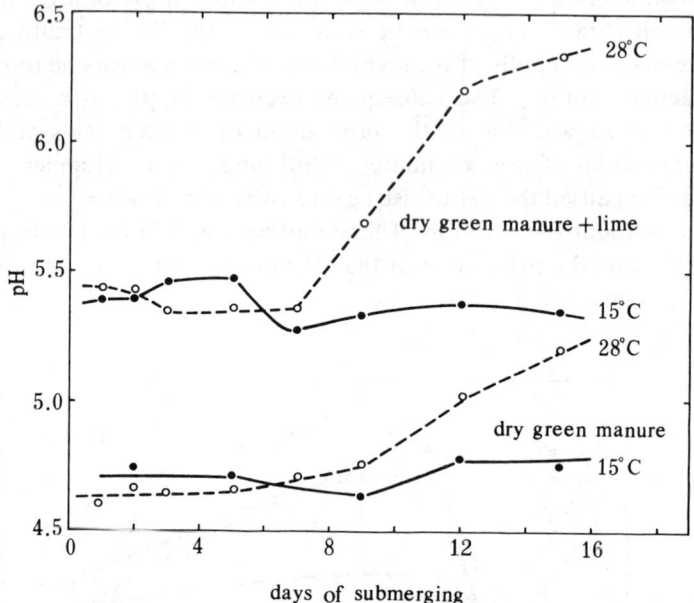

Fig. 7-12 Influence of temperature on change in pH of a red soil after submerging

7.2.2 Comparison between paddy soil and its parent soil

7.2.2.1 *Cultivated layer*

After the development of paddy soil from an acid parent soil, the pH and lime potential rise to a certain extent as a result of the increase in the amount of basic cations such as calcium and magnesium. The magnitude of the pH rise may be as high as 1.5 units or more (Fig. 7-13). The change in lime potential is even greater than that of the pH. For instance, the lime potential of a neutral paddy soil derived from yellow-brown soil is 5.4, about 1.5 units higher than its parent soil, and the lime potential of an acid paddy soil derived from red soil is about 5, higher than its parent soil by 2 units (Fig. 7-14). For a lateritic soil with a lime potential of 2.3, the lime potential increases by about 1 unit when the paddy soil is developed.

7.2.2.2 *Change in the profile*

Fig. 7-15 shows the variations of pH and lime potential within the profiles

7.2 Characteristics in the acidity of paddy soil

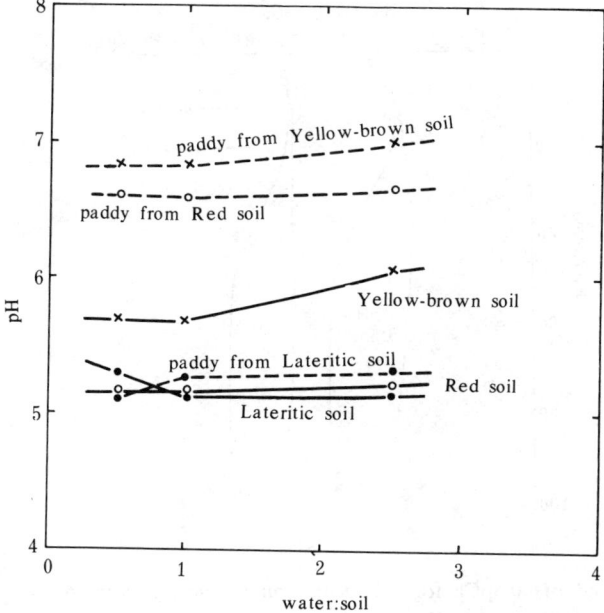

Fig. 7-13 pH of paddy soils and their parent soils

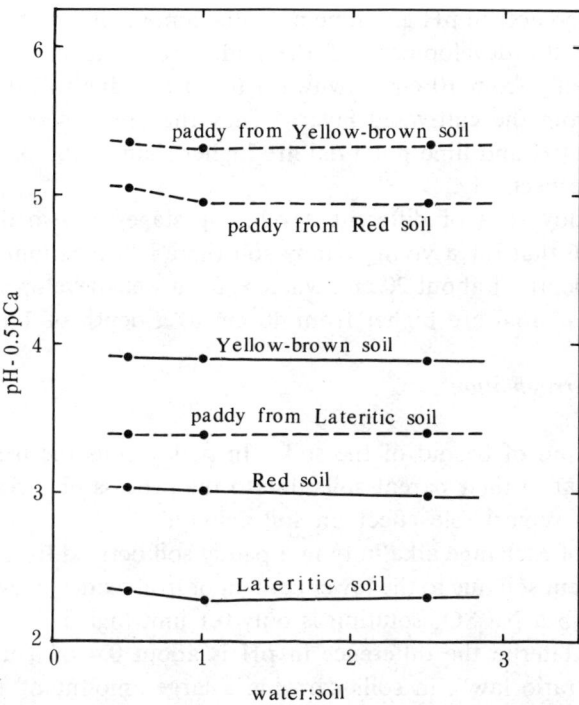

Fig. 7-14 pH-0.5pCa of paddy soils and their parent soils

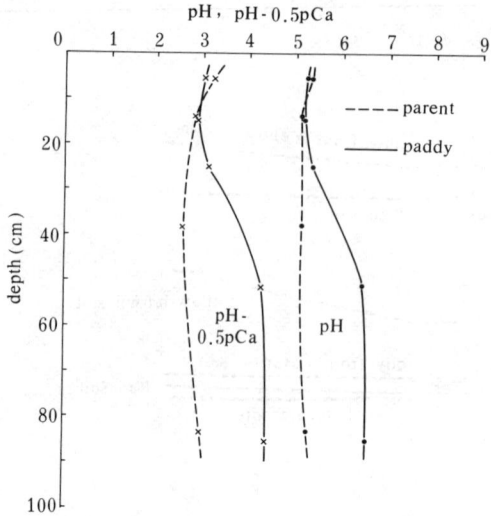

Fig. 7-15 pH and pH–0.5pCa for a lateritic soil profile and a paddy soil profile derived from it[9]

of a well-developed paddy soil derived from lateritic soil and its parent soil. For the lateritic soil, differences in pH and lime potential among different horizons are insignificant. After the development of the paddy soil, the pH and lime potential increase gradually from 10 cm downward to 50 cm, due to the illuviation of bases coming from the cultivated layer. They then remain constant to a depth of 90 cm. The pH and lime potential are higher than in the parent soil by 1.3 and 1.6 units respectively.

Soil acidity for paddy soils of different developing stages is also different. It is seen from Fig. 7-16 that for a young paddy soil there is a maximum of pH and lime potential at a depth of about 20 cm, whereas for a well-developed paddy soil the pH and lime potential are higher from 40 cm to a depth of 1 meter.

7.2.2.3 *Effect of deferrugination*

Iron oxides are a kind of basoid of the soil. In paddy soils the iron oxide content is lower than that in their parent soils due to the process of deferrugination. This may have a remarkable effect on soil acidity.

The lower amount of exchange alkalinity in a paddy soil derived from laterite as compared with its parent soil due to the lower content of iron oxides is evidenced in Fig. 7-17. The pH in a Na_2SO_4 solution is only 0.1 unit higher than that in water. For the parent laterite the difference in pH is about 0.4 of a unit.

According to the "ratio law", in soils carrying a large amount of negative charge and little positive charge the lime potential should be independent of the concentration of salts, and if the soil carries considerable positive charge the lime

7.2 Characteristics in the acidity of paddy soil

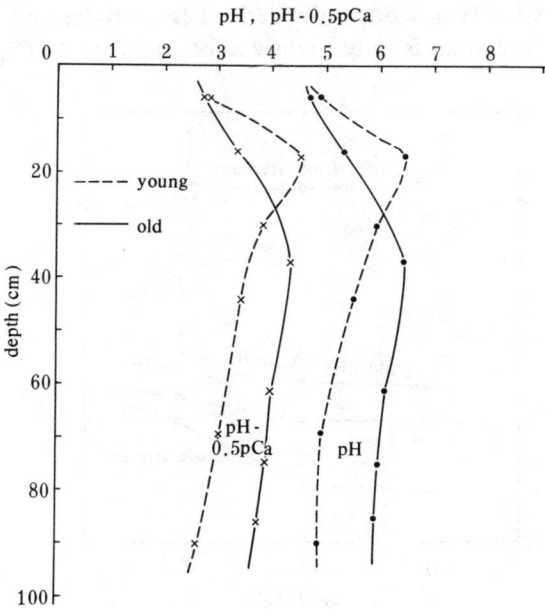

Fig. 7-16 pH and pH–0.5pCa of two paddy soil profiles at different developing stages[9]

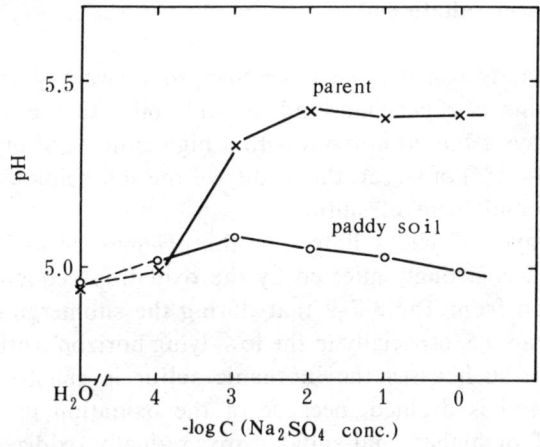

Fig. 7-17 Exchangeable alkalinity of a laterite and a paddy soil derived from it

potential should change with salt concentration. Fig. 7-18 shows the lime potential of two paddy soils and their parent soils in solutions of different concentrations of $CaCl_2$. For the red soil (containing 4.9% Fe_2O_3) and lateritic soil (containing 5.7% Fe_2O_3) carrying a certain amount of positive charge, the lime potentials are higher in concentrated $CaCl_2$ solution than in dilute solution by 0.2 and 0.6 unit respectively, whereas for two paddy soils derived from them and containing less iron oxides the lime potential is only slightly affected by the $CaCl_2$ concentration.

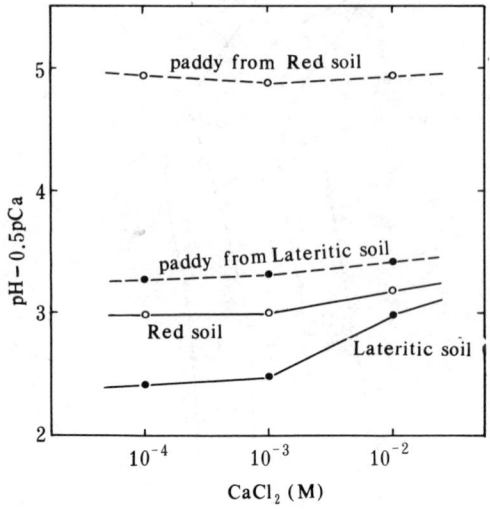

Fig. 7–18 Comparison of the effect of $CaCl_2$ concentration on pH–0.5pCa for two paddy soils and their parent soils[9]

7.2.3 Peculiarity of acid sulfate soil

The acid sulfate paddy soil derived from mangrove swampy soil along the seashores of South China is a peculiar kind of acid soil. In the lower part of such a soil there is always a buried horizon with a high content of organic matter and a high content (0.5—4%) of sulfur, the acidity of the soil being usually closely related to the amount and form of sulfur.

In the cultivated layer of acid sulfate soils the pH may be as low as 2.2 to 3.5 when dry. The pH is strongly affected by the oxidation–reduction status of the soil. It can be seen from Table 7–9 that during the submerging season the pH values are higher than 5.5, especially in the low-lying horizon with a strongly reducing condition. In such cases the inorganic sulfur is chiefly in the form of sulfide. Once the soil is drained, because of the oxidation in the soil, the Eh can rise to 400 mV or higher, and sulfides are gradually oxidized to sulfuric acid. In some cases the content of sulfuric acid may be 30 m.e. or more per 100 g of soil. Irrespective of whether such sulfuric acid remains in the form of free acid or subsequently reacts with mineral particles to form aluminum sulfate, the pH of the soil is low. However, the pH rises again if the soil is resubmerged. According to statistics[7], the correlation pattern of pH with the content of active acidity is very similar to that with the content of water-soluble sulfur (chiefly sulfates), implying that the change in soil pH is primarily determined by the change in active acidity, and the principal source of active acidity is aluminum sulfate.

TABLE 7-9 Influence of redox condition on pH of an acid sulfate soil (Dongguan)[7]

Horizon	Depth (cm)	Eh (mV)		pH		
		Submerged	Drained	Submerged	Drained	Air-dried
A	0—14	160	—	5.5	3.10	3.05
P	14—25	340	480	5.5	2.90	2.80
B_1	25—50	230	480	5.5	2.55	2.25
B_2	50—70	310	420	5.5	3.00	2.15
G	70—100	150	130	7.0	5.10	2.95

7.3 BUFFERING CAPACITY OF PADDY SOILS

Buffering capacity means the ability of a soil to retard the change in acidity after the addition of acidic or alkaline substances to the soil. The causes of buffering capacity in soils include: some weak acids such as carbonic acid, silicic acid, phosphoric acid and humic acid and their salts which in soil solution constitute buffering systems; various basic cations adsorbed on soil colloids which can buffer against H^+ ions and adsorbed H^+ which can buffer against OH^- ions, as well as some clay minerals which can act as a buffer through accepting protons of the solution. In acid soils with a pH of lower than 5 the presence of adsorbed aluminum ions makes it possible to neutralize the added OH^- ion by furnishing H^+ from, for instance, $Al(OH)_6^{3+}$:

$$2Al(H_2O)_6^{3+} + 2OH^- \longrightarrow Al_2(OH)_2(H_2O)_8^{4+} + 4H_2O$$

In the following, the relationship between buffering capacity and some soil factors will be briefly discussed.

7.3.1 Buffering capacity in different horizons

In different horizons of a profile the kind and amount of buffering substances may be different, and so may the buffering capacity. The buffering capacity of humus is larger by ten to one hundred times than the mineral part of the soil, and therefore in the cultivated layer which is high in organic matter content the buffering capacity is larger than that in the lower horizons. In Fig. 7–19 are shown acid–base titration curves for a paddy soil derived from Quaternary red clay. The buffering capacity for low–lying horizons is smaller than that for the cultivated layer, while the glei horizon is the smallest. If the pH span after the addition of 2 m.e. of HCl and 2 m.e. of NaOH per 100 g of soil is taken as a relative index of buffering capacity of the soil, it was observed that the buffering capacity of the four horizons from the cultivated layer downward was 2.7, 2.9, 4.4 and 5.4 pH units respectively. It will also be observed from Table 7–10 that the buffering power is strongest in the cultivated layer and weakest in the

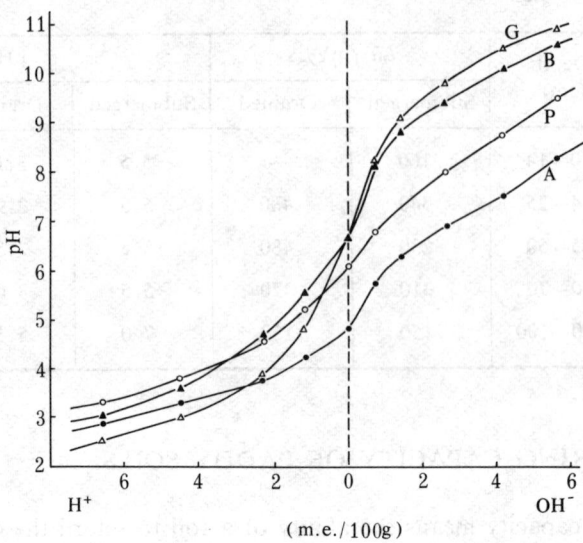

Fig. 7-19 pH titration curves for different horizons of a paddy soil profile derived from red soil (Zhejiang)

TABLE 7-10 Buffering capacity of different horizons of some paddy profiles[5]

Locality	Horizon	Depth (cm)	pH change caused by 2 m.e./100g of		
			HCl	NaOH	Total
Hengyang	A	0—12	0.7	1.4	2.1
	P	12—23	0.9	1.9	2.8
	B_1	23—33	1.1	2.4	3.5
	B_2	33—64	0.8	1.9	2.7
Hengyang	A	0—16	0.9	2.0	2.9
	P	16—30	1.6	1.5	3.1
	B_1	30—60	1.5	2.0	3.5
	B_2	60—66	2.0	1.3	3.3
	C	66—	2.0	1.4	3.4
Dongxiang	A	0—10	1.3	1.5	2.8
	P	10—15	1.4	2.0	3.4
	B	15—52	1.1	1.9	3.0
	G	52—58	1.9	2.6	4.5

7.3 Buffering capacity of paddy soils   153

glei horizon, which is in conformity with the trend shown in Fig. 7–19. The cause of the low buffering capacity in the glei horizon lies probably in the low cation–exchange capacity of the soil caused by mechanical leaching and chemical destruction of clay particles as a result of reductive eluviation.

7.3.2 Buffering capacity of different types of soil

Fig. 7–20 shows the buffering curves for paddy soils derived from different parent materials. The order of buffering power against acid for the three materials is: weathering product of limestone (soil 1) > weathering product of granite (soil 3) > Quaternary red clay (soil 2), and that against alkali is: weathering product of limestone > weathering product of granite ≈ Quaternary red clay. The large buffering capacity of the paddy soil derived from limestone is due to the heavy texture and the high content (7%) of organic matter.

Fig. 7–21 shows a comparison in buffering curve among paddy soil, upland soil and wasteland soil of the same soil type. The larger buffering capacity of the paddy soil is evidenced by the smoothness in the form of the curve as compared with other two soils, due probably to its large cation-exchange capacity, high base-saturation percentage and high pH.

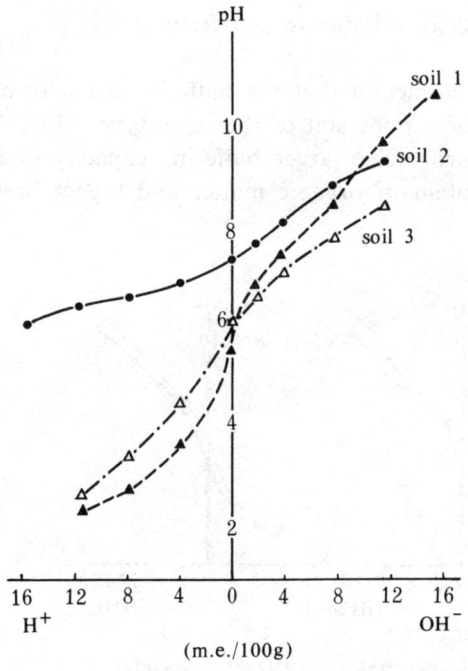

Fig. 7–20 Buffering curves of different types of paddy soil[1]

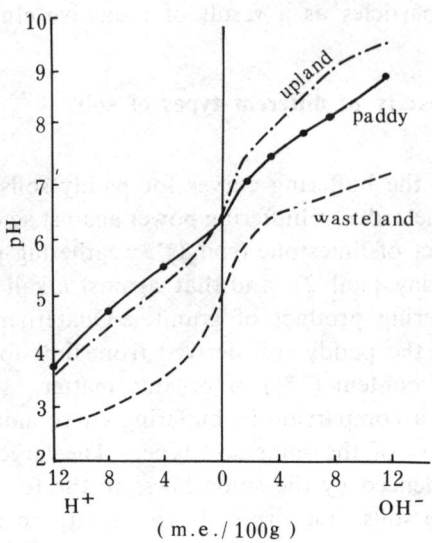

Fig. 7-21 Buffering curves of paddy soil, upland soil and wasteland soil of the same parent material (purplish soil)[6]

7.3.3 Buffering capacity in relation to soil fertility

It is a common phenomenon that the buffering capacity of a fertile paddy soil is larger than that of a poor soil of the same type. Fig. 7-22 shows some comparisons in this regard. The larger buffering capacity of fertile paddy soil is due to the higher content of organic matter and higher base-saturation percentage.

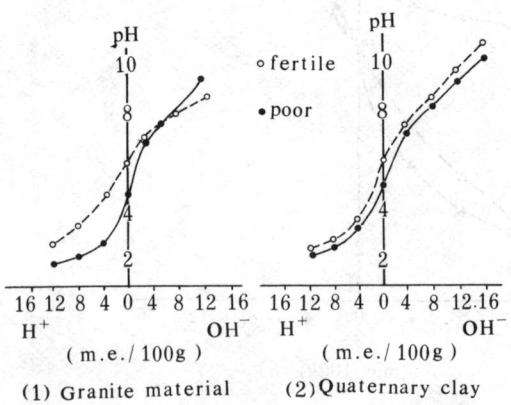

(1) Granite material   (2) Quaternary clay

Fig. 7-22 Buffering curves of paddy soils with different fertility[1]

7.4 ACIDITY REGIME OF PRINCIPAL TYPES OF PADDY SOIL

The acidity regime of a paddy soil inherits the acidity property from its parent soil, and is also influenced by agricultural measures. According to the pH value, paddy soils may be distinguished as three main types, namely acid, neutral and calcareous. A principal characteristic feature of paddy soil with respect to acidity is that after submerging the reaction of these three types of soil all tends to approach to neutral, with a steady pH of 6.5 to 7.5 in most cases.

Acid paddy soils are distributed in the southern part of China. Because a large part of these soils are derived from strongly leached red or yellow soils, they are frequently called paddy soil derived from red soil, paddy soil derived from yellow soil, etc. In this region the leaching of bases from the paddy soil itself is also strong, hence the soil is generally base–unsaturated. According to a large number of determinations, paddy soils derived from the weathering product of granite in Guangdong, Fujian and Jiangxi provinces usually have a pH of 5.0 to 6.0. For paddy soils derived from Quaternary red clay in Jiangxi, Hunan, Zhejiang and Guangxi provinces the pH varies from 5.0 to 7.0, relating to the intensity of cultivation and manuring system of the region, and it is generally higher for fertile soils or if liming is included in the manuring practice. Paddy soils derived from river alluvium in South China are usually also acid in reaction, the actual pH varying with the source and period of deposit.

Neutral paddy soils are mainly distributed in the middle and lower reaches of the Changjiang River (Yangtze River), and have a pH of 6.5—7.5. The principal parent materials are yellow–brown soil, old river alluvium and old lacustrine deposits. Some paddy soils derived from purplish soils or limestone soils in South China are also neutral in reaction.

Most of the calcareous paddy soils are distributed in North China. The parent soils are calcareous due to the weakness in leaching of bases, and paddy soils developed on them are also invariably calcareous. The pH of these soils is around 8. Some paddy soils derived from alkali soils may have a pH of 9 or higher. In South China some paddy soils derived from the weathering products of limestone or purplish soil may also be calcareous. In the lower reaches of the Changjiang River some young paddy soils derived from recent alluvium also contain a certain amount of calcium carbonate.

Viewing the geographical distribution of paddy soils in China as a whole, the general pattern is that the paddy soils in the north and northwest part are calcareous, the majority of those in Central China are neutral, and those in South China are mostly acid. Hence there is a general regularity of increase in soil pH from South China northward. However, exceptions are not uncommon. For example, in Northeast China there are paddy soils having a neutral or even weakly acid reaction, and in South and Central China there are young paddy soils containing variable amount of carbonates. One special case is that in some parts of South China the soil may contain lime concretions due to long–term overliming.

REFERENCES

(1) Institute of Soil Science, 1961. Soil Environment of High-yield Rice. Chapter 5. Science Press, Beijing.
(2) Yu Tian-ren et al., 1976. Electrochemical Properties of Soils and Their Research Methods. (revised ed.) Chapters 9, 10. Science Press, Beijing.
(3) Institute of Soil Science, 1978. Soils of China. Part B, Chapter 9. Science Press, Beijing.
(4) Ling Yun-xiao and Yu Tian-ren, 1957. Soil acidity in relation to exchangeable hydrogen and aluminum. Acta Pedologica, **5**: 234–244.
(5) Yu Tian-ren and Ding Chang-pu, 1958. On the status of exchangeable bases and its relation to the genesis of paddy soils derived from red soils. Soils Bulletin, **33**: 31–43.
(6) General Soil Survey Group, 1959. On the genesis and classification of paddy soils of South China. Acta Pedologica, **7**: 28–41.
(7) Gong Zi-tong and Zhou Rui-yung, 1964. Genesis of acid sulfate soils. Acta Pedologica, **12**: 183–190.
(8) Guizhou Soil Survey Team, 1980. Soil resources and land utilization of Malu Commune, Changshun County, Guizhou Province. Soils Bulletin, **37**: 145–160.
(9) Wang Ching-hua and Yu Tian-ren, 1981. Lime potential of soils as directly measured with two ion-selective electrodes. Z. Pflanzenerhähr. Bodenkunde, **144**: 514–523.
(10) Cang Dong-qing and Yu Tian-ren, 1981. Determination of pH of paddy soils in situ. in "Proc. Symp. Paddy soil". pp. 709–715. Science Press, Beijing (in English).

CHAPTER 8

ELECTRICAL CONDUCTIVITY

WU JUN, SUN HUI-ZHEN AND ZHANG DAO-MING

Electrical conductivity is the phenomenon of a transfer of electricity carried by charged particles (ions, colloids) under the force of an applied electric field. The electrical conductivity of paddy soil follows the general behavior of the electrical conductivity of soils. However, there are also characteristic features peculiar to the paddy soil. These are partly formed during the long-term genetic process of paddy soil, and are partly caused by the short-term influence of agricultural measures.

This chapter deals with the factors affecting the electrical conductivity of soils, the characteristics in the electrical conductivity of paddy soil and the relationship between electrical conductivity and soil fertility, and finally, the electrical conductivity of principal types of paddy soil.

8.1 FACTORS AFFECTING THE ELECTRICAL CONDUCTIVITY OF SOILS

The electrical conductivity of a soil is determined chiefly by the amount and nature of charged particles, that is the quantity of electric charge carried by soil clays and the kind and quantity of ions, and is also affected by some geometric factors. For soils carrying an insignificant amount of soluble salts and under conditions where geometrical factors are kept constant, the electrical conductivity is determined primarily by the nature of the clay and its interactions with ions. These will be discussed in the following sections.

8.1.1 Interactions between clay and ions

The migration velocity of clay particles in an electrical field is in the same order of magnitude as that of ions[2]. However, the quantity of electric charge carried by different clays differs greatly (cf. Chapter 6). The conductivity, i.e., the mobility of ions in the soil, varies with their position in the electrical double-layer. It is generally assumed that the behavior of ions in the diffused Gouy layer with respect to conducting electricity is not different from ions in the bulk solution, and in the innermost one or two ion layers of the Stern layer adjoining to the clay surface the ion does not participate in electrical conductance, whereas the mobility of ions in other parts of the Stern layer is only several to twenty or thirty percent of the normal mobility of the ion. Because the distribution of adsorbed ions in the electrical double-layer is affected by a variety of factors, the conductivity of various soils under different conditions differs greatly.

Since it is impossible to distinguish the difference in conductivity of ions of various parts of the electrical double-layer by conventional conductivity methods, we can only evaluate the overall interaction between adsorbed ions and clay particles based on the difference in conductivity under various conditions. Fig. 8-1 shows the change in conductivity of a hydrogen- and aluminum-saturated paddy soil derived from yellow-brown soil and its parent soil when titrated with KOH. In interpreting this kind of titration curve the soil may be regarded as a weak acid. The low electrical conductivity of the H,Al-saturated soil is due to the low degree of dissociation of ions from the clay. Following the addition of KOH, the conductivity increases steadily. This is because the dissociation of adsorbed K^+ ions is easier than hydrogen or aluminum ions. When the added KOH exceeds a certain limit the conductivity increases with a steeper slope due to the enhancement in the dissociation of adsorbed K^+ ions and especially the presence of free OH^- ions. The shape of the whole titration curve is similar to that of a weak acid titrated with a strong base, and it is possible to draw two straight lines from the curve. The intersection point of the lines is equivalent to the cation-exchange capacity of the soil, just like the concentration of a weak acid. It can be seen from the figure that the curves of the paddy soil and its parent soil are very similar, and the intersection points for the two soils correspond to 18.9 and 17.3 m.e. per 100 g of soil respectively. This kind of extrapolation is the theoretical basis for estimating the cation-exchange capacity of a soil based on the conductometric titration curve. The shape of the curve shown in Fig. 8-1 is representative of a large number of paddy soils.

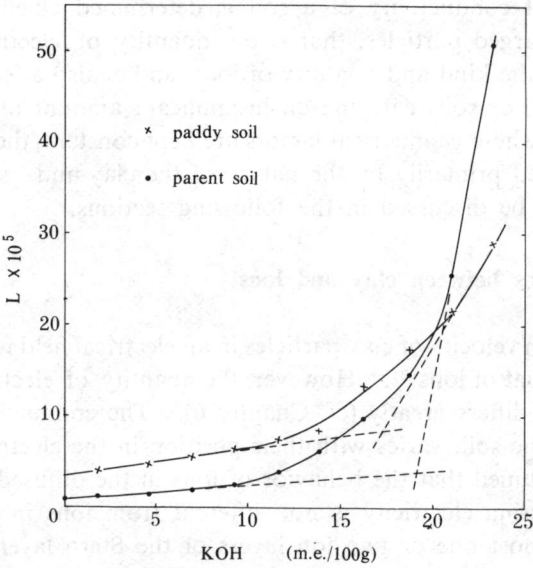

Fig. 8-1 Conductometric titration curves of yellow-brown soil and paddy soil derived from it (courtesy of A. Z. Zhao)

8.1 Factors affecting the electrical conductivity of soils

Fig. 8-2 shows conductometric titration curves for a laterite and a paddy soil derived from it. The curves differ from those of Fig. 8-1 in that it is very difficult to distinguish two straight lines with different slopes. This means that in the titration curve there is no threshold region beyond which the conductivity increases sharply. It must be assumed that this is a reflection of the peculiarity in the surface property of clay in the paddy soil derived from laterite. It should also be noticed that this characteristic feature is even more pronounced for the parent laterite.

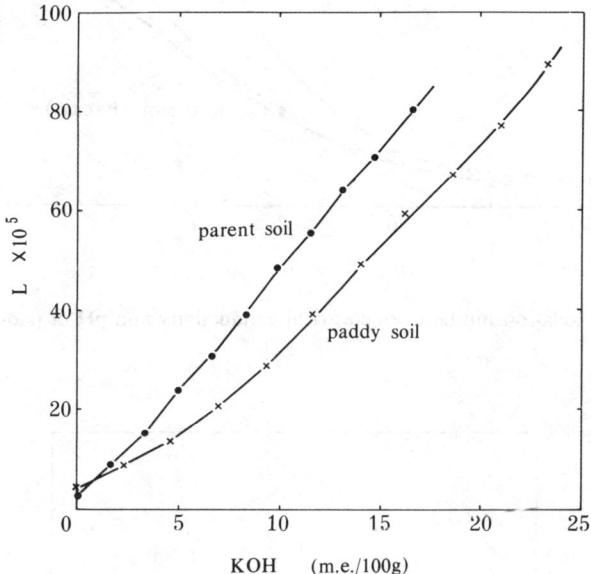

Fig. 8-2 Conductometric titration curves of laterite and paddy soil derived from it (courtesy of A. Z. Zhao)

Because in natural strongly acid soils the exchangeable cations are mostly hydrogen and aluminum, the change in conductivity is very similar to those shown in Fig. 8-1 if titrated with a base. The results for three strongly acid paddy soils are shown in Fig. 8-3. Following the addition of NaOH, the pH increases, and so does the conductivity.

The degree of dissociation for different adsorbed cations is different. It can be seen from a comparison shown in Fig. 8-4 that at a same pH the specific conductance of a paddy soil containing calcium as the dominant cation is much lower than that of the soil containing sodium as the dominant cation.

8.1.2 Physical factors

In addition to the influence of charged clay particles on the mobility of ions when they migrate in an applied electric field, the migration path of ions in soil

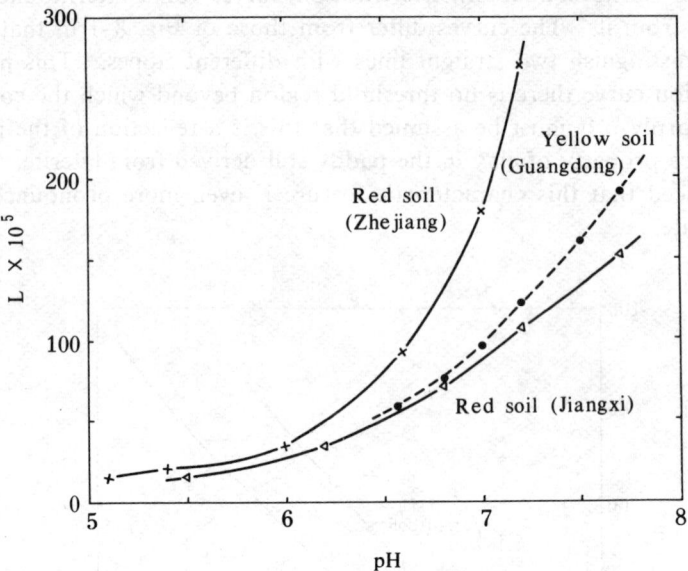

Fig. 8-3　Relationship between electrical conductivity and pH of paddy soil

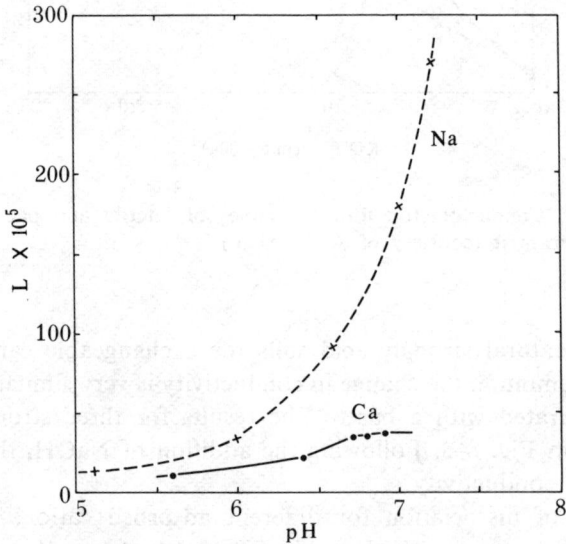

Fig. 8-4　Effect of kind of adsorbed cation on electrical conductivity of the soil (soil pH adjusted with Na_2CO_3 or $CaCO_3$)

is also different from that in free solution. For instance, when the water content of the soil is low there may be water as well as air in the pores. In such a case the effective volume that conducts electricity would be smaller than that of the soil

8.1 Factors affecting the electrical conductivity of soils

in which the pores are completely filled with water, and hence its conductivity would be lower. Therefore, within a certain water content range the conductivity of a soil will increase with the increase in water content (Fig. 8–5). When the water content is sufficiently high, the conductivity is hardly affected by the variation in water content within the range of 70—130% of the water–holding capacity. If the water content is further increased, the conductivity decreases due to the dilution effect. Table 8–1 shows the change in conductivity with water content for some paddy soils.

During the rice–growing period, the water content in the cultivated layer

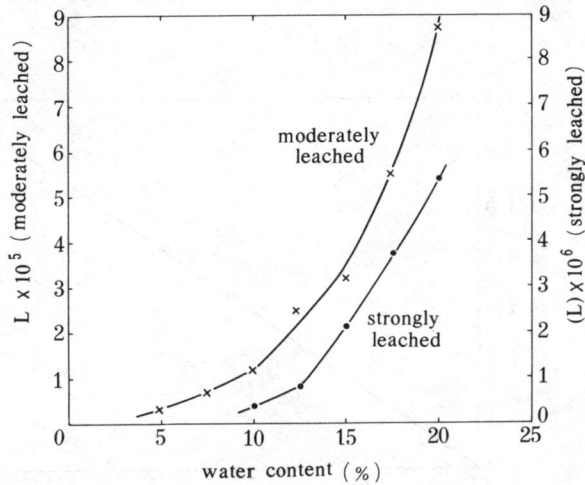

Fig. 8–5 Relationship between electrical conductivity and water content of the soil at low water content

TABLE 8–1 Effect of water content on electrical conductivity of paddy soils

Soil	Water–holding capacity (%)	$L \times 10^5$				
		50%*	70%	100%	130%	1:1**
Chingge	43.7	—	33	34	34	34
Huangge	45.1	13	31	34	33	31
Beisha	57.7	1.5	17	22	22	21
Wusha	61.7	4.4	26	28	28	27
Wuni	63.1	5.8	30	38	34	34
Wuni	64.2	6.8	54	54	52	49
Huangsheng	85.1	6.2	32	39	39	—

* As % of water–holding capacity
** Water to soil ratio

is generally in the range of 85—120% of the water-holding capacity of the soil. So for the measurement of the electrical conductivity of paddy soils in situ it is not very important to consider the effect of the variation of water content.

When the pores are completely filled with water, the ratio of solid phase to liquid phase will also affect the conductivity of the soil. Fig. 8-6 shows the decrease in conductivity when three paddy soils are mixed with varying amounts of quartz sand with insignificant conductivity. In constructing the curve, the reciprocal of the relative conductivity, i.e., the relative resistivity, is used as the unit, taking the value of the original soil without the addition of sand as 1. It is seen that the resistivity of the soil increases linearly with the increase in sand content.

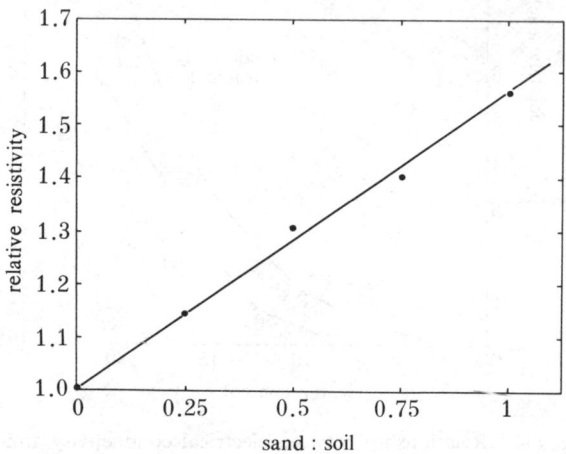

Fig. 8-6 Influence of sand addition on electrical conductivity of the soil (water-saturated) (mean value of three soils)

In addition to the geometric factor mentioned above, soil texture will affect the conductivity through differences in electric charge carried by clay particles and compensating ions. It was observed in Taiho district of Jiangxi Province that the specific conductance of the cultivated layer of a loamy soil was 2.12×10^{-5} mho/cm, whereas for a clayey soil of the same parent material and a similar fertility level it was 7.68×10^{-5} mho/cm.

Under conditions of similar texture, the compactness will also have some influence on the conductivity of the soil. It was observed that when the water content was below the water-holding capacity, the conductivity increased with the compactness of the soil (Fig. 8-7), due apparently to the decrease in the volume fraction occupied by soil air.

8.1.3 Soluble salts

If the soil contains soluble salts in addition to exchangeable cations, the surplus soluble ions will cause the conductivity of the soil to increase. Fig 8-8

8.1 Factors affecting the electrical conductivity of soils

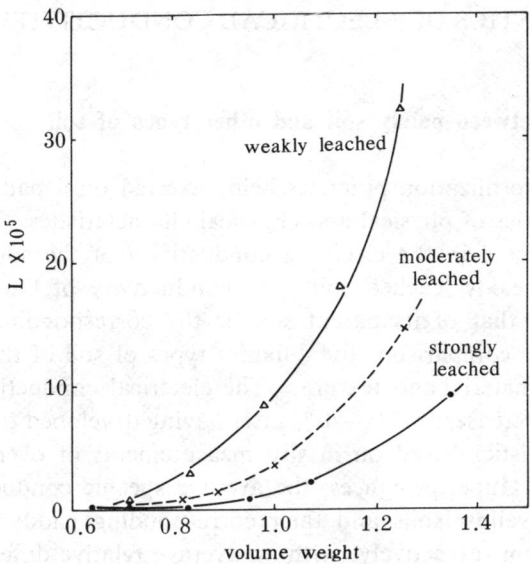

Fig. 8-7 Relationship between electrical conductivity and volume weight of the soil (water content 15%)

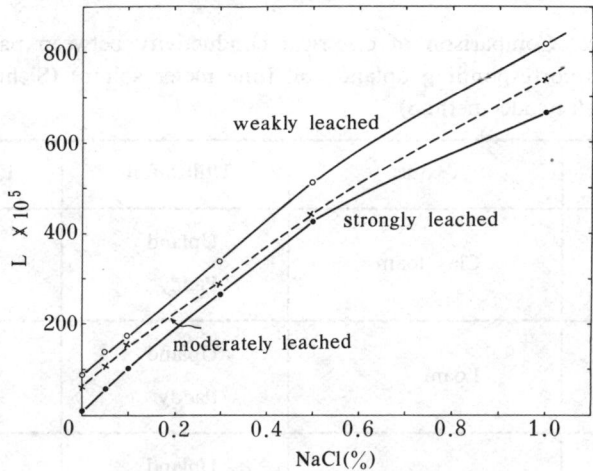

Fig. 8-8 Effect of NaCl on electrical conductivity of paddy soils

shows some examples in this regard for three paddy soils. The electrical conductivities of three soils were in the following order: strongly leached paddy soil (Zhejiang)<moderately leached (Jiangsu)<weakly leached (Tianjin). After the addition of NaCl, the conductivity increased nearly linearly with the increase in the amount added. This phenomenon is the theoretical basis for estimating the soluble salt content of salt-affected soils.

8.2 CHARACTERISTICS OF ELECTRICAL CONDUCTIVITY OF PADDY SOIL

8.2.1 Comparison between paddy soil and other types of soil

Many irrigation–fertilization practices being exerted on a paddy soil during its genesis lend it a series of physical and chemical characteristics. These characteristics must be reflected in the electrical conductivity of the soil. Except for those derived from weakly leached soils, the conductivity of the paddy soil is generally higher than that of its parent soil or the corresponding upland soil. Table 8–2 shows some comparisons for different types of soil of the same region and of similar parent material and texture. The electrical conductivity of yellow soil or purplish soil increases by 20—70% after having developed to a paddy soil. According to the statistics based on in situ measurements of over five hundred fields in Sichuan and Hubei provinces, the average specific conductances of the one–meter solum of yellow soils and their corresponding paddy soils are 19.2 and 31.6×10^{-5} mho/cm respectively, with an average relative difference of 65%. For purplish soils the average relative difference is 33%, namely the electrical conductance increases from 52.6 to 70.2×10^{-5} mho/cm when the soil has developed to paddy soil. Of course, the water content in the lower part of the solum of paddy soils may be slightly higher in some cases than that in upland soils even

TABLE 8–2 Comparison of electrical conductivity between paddy soil and its corresponding upland soil (one meter solum) (Sichuan, Hubei) (4–electrode method)

Soil type	Texture	Utilization	$L \times 10^5$
Yellow soil	Clay loam	Upland	17.1
		Paddy	26.1
Yellow soil	Loam	Upland	34.2
		Paddy	50.5
Purplish soil	Sandy loam	Upland	20.9
		Paddy	35.1
Purplish soil	Loam	Wasteland	37.4
		Paddy	64.9
Purplish soil	Loam	Upland	28.9
		Paddy	45.5

in the season when upland crops are grown, and this will influence the conductivity of the soil. Nevertheless, it must be admitted that differences in the kind and content of ions between the two types of soils should be an important or the dominant factor in accounting for the differences in electrical conductivity, since this difference in conductivity also exists between soils of the cultivated layer. For example, in Lanxi district of Zhejiang Province, the specific conductance of the cultivated layer of an upland red soil is 1.33×10^{-5} mho/cm, whereas for an adjoining paddy field it is 2.17×10^{-5} mho/cm. In Taiho district of Jiangxi Province, the respective figures of 1.35 and 11.2×10^{-5} mho/cm have been observed.

The increase in conductivity during the genetic process of paddy soil is the result of a variety of factors among which the input of mineral elements introduced through fertilizing and manuring is usually the most important. Irrigation water also contains some mineral materials. Besides, clay particles carried in by farm manure and irrigation water may also influence the conductivity of the paddy soil.

8.2.2 Change in electrical conductivity during submergence

The electrical conductivity of paddy soils increases after submerging. This increase does not imply direct effect of water content of the soil as mentioned in the last sections, but denotes the result of chemical changes occurring in the soil. Generally, in the first stage of submergence the electrical conductivity increases sharply; then the rate of increase tends to lessen gradually until a steady value is attained. Fig. 8–9 shows some representative examples. Tables 8–3, 8–4 and 8–5 show the increase in conductivity of some paddy soils (leached strongly, moderately and weakly respectively) after submergence for 21 days.

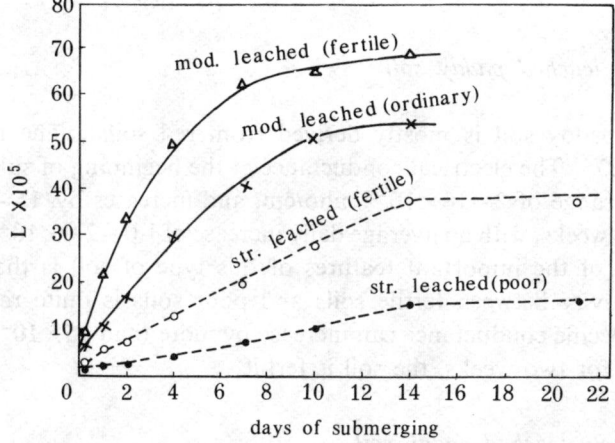

Fig. 8–9 Change in electrical conductivity of paddy soils during submergence[3]

TABLE 8-3 Change in electrical conductivity of strongly leached paddy soils after submerging[3]

Locality	Fertility level	pH	O.M. (%)	$L \times 10^5$			
				Initial	22nd day	Increase	Increase / day
Zixi	Low	5.3	2.72	8.6	8.6	0	0
	High	6.6	3.10	11.9	38.4	26.5	1.3
Yichun	Low	6.1	2.68	3.0	20.0	17.0	0.8
	High	6.9	3.35	6.9	47.6	40.7	1.9
Changsha	Low	5.4	2.06	2.0	16.1	14.1	0.7
	High	6.3	2.70	3.2	37.4	34.2	1.7
Luoping	Low	6.9	2.32	9.0	33.9	24.9	1.1
	High	6.1	4.02	6.3	36.4	30.1	1.4
Anyi	Low	5.9	2.62	2.1	21.8	20.7	1.0
	High	—	2.69	5.9	65.1	59.2	2.8
Pingxiang	Low	6.6	4.09	4.7	35.1	30.4	1.4
	High	5.8	4.47	10.5	47.2	36.7	1.7
Yiwu	Plowpan	6.8	0.70	1.6	7.0	5.4	0.3
	Surface	5.2	1.81	5.6	27.5	21.9	1.0

8.2.2.1 *Strongly leached paddy soil*

This type of paddy soil is mostly derived from red soils. The pH of the soil is usually 5—6.5. The electrical conductance at the beginning of submergence is generally in the range of 2—10×10^{-5} mho/cm, and increases by 15—35×10^{-5} mho/cm after three weeks, with an average daily increase of 1.0—2.0×10^{-5} mho/cm (Table 8-3). One of the important features of this type of soil is that the difference in conductivity between fertile soils and poor soils is quite remarkable. Generally, if the specific conductance can increase by more than 25×10^{-5} mho/cm after submergence for two weeks, the soil is fertile.

8.2.2.2 *Moderately leached paddy soil*

This type of soil is usually neutral in reaction. The initial conductivity

8.2 Characteristics of electrical conductivity of paddy soil

TABLE 8–4 Change in electrical conductivity of moderately leached paddy soils after submerging[3]

Locality	Fertility level	pH	O.M. (%)	$L \times 10^5$			
				Initial	22nd day	Increase	Increase / day
Sungjiang	Ordinary	—	—	6.9	50.5	43.6	2.1
	High	—	—	8.8	64.1	55.3	2.6
Wuxi	Ordinary	5.6	2.11	3.1	11.0	7.9	0.4
	High	6.1	2.23	4.8	13.2	8.4	0.4
Nanjing	Ordinary	6.8	2.20	6.7	23.2	16.5	0.8
	High	6.4	2.27	8.8	18.1	9.3	0.4
Xiaogan	Ordinary	5.5	2.35	12.4	38.8	26.4	1.3
	High	5.5	1.89	7.9	33.8	25.9	1.2
Xiaogan	Ordinary	5.4	2.41	5.3	24.3	19.0	0.9
	High	5.4	2.29	6.5	25.2	18.7	0.9

is in the range of $5\text{—}15 \times 10^{-5}$ mho/cm, and the increase after three weeks of submergence varies from 10 to 55×10^{-5} mho/cm. For this type of soil the relationship between electrical conductivity and fertility level is usually not distinct. However, in some cases the difference between fertile and poor soils is still noticeable.

8.2.2.3 Weakly leached paddy soil

This type of soil is derived from calcareous or salt-affected soils, or soil which has been overlimed. The initial specific conductance is usually in the range of $6\text{—}20 \times 10^{-5}$ mho/cm, but in some cases it may be 40×10^{-5} mho/cm. The electrical conductivity increases by $25\text{—}40 \times 10^{-5}$ mho/cm after submergence for three weeks, and in special cases the increase may be as high as 60×10^{-5} mho/cm or more.

8.2.2.4 Causes of increases in electrical conductivity

The increase in electrical conductivity after submerging is related to the increase in amounts of ammonium, ferrous and manganous ions and the destruction of the mineral part of the soil. However, the principal cause of this increase is the decomposition of organic matter. The data in Tables 8–3, 8–4

TABLE 8-5 Change in electrical conductivity of weakly leached paddy soils after submerging[3]

Locality	Description	pH	O.M. (%)	L × 10⁵			
				Initial	22nd day	Increase	Increase/day
Luocheng	Poor	7.7	2.52	7.4	31.9	24.5	1.1
	Ordinary	7.9	4.11	15.4	52.9	37.5	1.8
	Fertile	7.9	6.96	21.2	64.9	43.7	2.1
Changshu	Gleyed	7.7	9.40	47.4	80.4	33.0	1.6
	Fertile	8.0	3.99	19.4	49.0	29.6	1.4
Nanjing	Ordinary	8.0	3.05	9.5	34.7	25.2	1.2
	Fertile	8.1	3.51	10.7	37.7	27.0	1.3
Ganzhou	Fertile	6.9	3.85	7.0	48.9	41.9	2.0

and 8-5 show that for the same type of soil there is a definite relationship between the increase in conductivity during submergence and the organic matter content of the soil. It can be seen from Fig. 8-10 that when submerged at 28°C

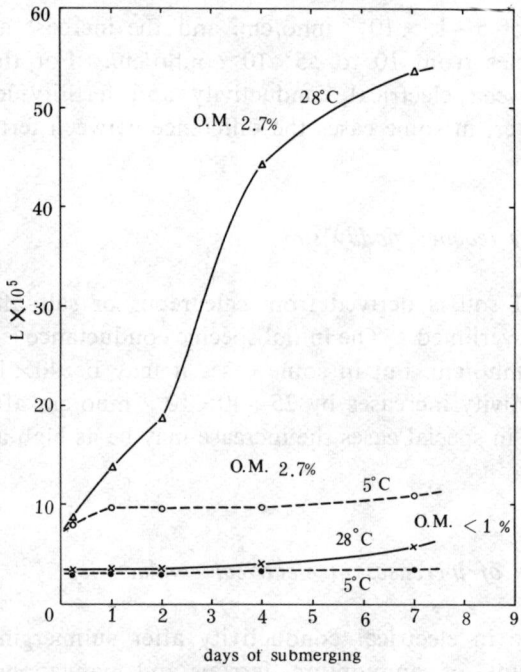

Fig. 8-10 Relationship between increase in electrical conductivity and decomposition of organic matter[3]

the increase in conductivity is larger for the soil containing more organic matter; at 5°C the conductivity increases only slightly due to the weakening of the microbiological decomposition of organic matter. It has also been observed that after the removal of most of the organic matter from the soil the increase in electrical conductivity during submergence is very small. This suggests that the increase in electrical conductivity during submergence is due chiefly to the decomposition of organic matter. However, the increase in electrical conductivity is not linearly proportional to the organic matter content of the soil, and there are also cases in which the increase in conductivity is not large even when the organic matter content of the soil is high. One of the most important reasons for the complexity of this relationship lies in the different resistance to decomposition of the different organic substances.

The periodical change in electrical conductivity following the alternation of submerged and drained seasons of the paddy soil is a universal phenomenon. When comparisons of electrical conductivity are made between different soils, it is necessary to consider this special case.

8.3 ELECTRICAL CONDUCTIVITY IN RELATION TO THE FERTILITY STATUS OF PADDY SOIL

8.3.1 Relationship between electrical conductivity and fertility level of the soil

Although conductivity can only reflect the overall ion status of a soil solution and can not distinguish nutrients from other ions, it can frequently be observed that there is a close relationship between electrical conductivity and the fertility level of non-salined paddy soils. This is because in farm manure which is the chief fertilizer for rice in China the range of variation in relative proportions of various ions is not large, and therefore the difference in total concentration of ions between two soils can roughly reflect the difference in nutrient level. Table 8–6 shows some examples in this regard. For these acid paddy soils the

TABLE 8–6 Relationship between electrical conductivity and fertility level of paddy soil (Jiangxi) (2–electrode method)

Parent soil	Fertility level	$L \times 10^5$
Yellow soil	Ordinary	16.3
	High	24.7
Alluvium	Ordinary	26.7
	High	32.3
Red soil	Ordinary	37.7
	High	46.3
Red clay	Ordinary	22.4
	High	35.5

electrical conductivity of the cultivated layer of fertile soils in the rice–growing period is higher by tens percent than the corresponding soils with medium fertility level. Another set of measurements made in Jinhua district of Zhejiang Province showed that the specific conductance of the 0—30 cm layer of a fertile paddy soil derived from red soil was 12.7×10^{-5} mho / cm, whereas the corresponding figures for the medium fertility level, poor and very poor soils derived from similar parent soils in the same region were 10.4, 6.97 and 3.39×10^{-5} mho / cm respectively. In Taiho district of Jiangxi Province, a similar tendency has also been observed for soils derived from red soils (Table 8–7).

TABLE 8–7 Relationship between electrical conductivity of the cultivated layer and fertility level of paddy soil (Taiho) (4–electrode method)

Fertility level	Usual yield of rice (kg / ha)	$L \times 10^5$
Lowest	3000	5.37
Lower	4000	10.8
Medium	6000	15.1
High	8000	28.3

All of the data mentioned above show that for acid paddy soils the correlation between electrical conductivity and the fertility level of the soil is quite clear.

The reason for this close correlation in acid paddy soils may be explained as follows: Because there is an insignificant amount of soluble salt, the conductivity of the soil is caused mainly by exchangeable ions such as calcium, magnesium, potassium and ammonium. Calcium and magnesium are important nutrients for the growth of rice. There is generally a definite correlation between the amount of calcium and magnesium ions and that of other nutrient ions. And, as compared with other types of soil, the proportions of potassium and ammonium in the total ions are larger. Therefore, for acid soils the specific conductance can be used as a single index for comprehensively characterizing the fertility level of the soil.

8.3.2 Effect of agricultural measures on electrical conductivity

8.3.2.1 *Fertilization*

The kind and amount of fertilizer can remarkably affect the conductivity of the soil. For example, in a paddy field of the mountain region of Jiangxi Province, the specific conductances were 44.2, 52.1 and 61.0×10^{-5} mho / cm respectively when 7.5, 20 and 30 ton / ha of pig dung were applied. In an al-

8.3 Electrical conductivity in relation to fertility

luvial field of Xiaogan district of Hubei Province, the specific conductances were 52.1, 66.2 and 71.3 × 10^{-5} mho / cm respectively in plots where 85, 130 and 210 ton / ha of farm manure had been applied[1]. The application of 380 kg / ha of superphosphate led to an increase in specific conductance from 51.1 to 57.2 × 10^{-5} mho / cm. Topdressing also had a clear effect on the conductivity of the soil. For instance, in an acid paddy soil a topdressing of 75 kg / ha of ammonium nitrate induced the specific conductance of the soil to increase from 37.6 to 60.9 × 10^{-5} mho / cm, i.e., by about 60%. Even in paddy soils with a relatively high fertility level, e.g. the "White soil" fields of Taihu region, it has been observed that the application of 300 and 600 kg / ha of nitrogen fertilizer led to an increase of specific conductance from the original value of 49.2 to 78.6 and 97.0 × 10^{-5} mho / cm respectively. For more fertile paddy soils, the effect of fertilization can still be observed (see Table 8–8).

TABLE 8–8 Effect of fertilization on electrical conductivity of the cultivated layer of paddy soil (Wuxi)

Locality	Ammonium bicarbonate (kg / ha)	L × 10^5	
		Before fertilization	After fertilization
Yiuyi	300	56.3	80.4
Dongting	500	67.0	97.0
Shuguang	500	55.6	94.6
Wangting	250	55.9	72.7
		Mean 58.7	Mean 86.2

8.3.2.2 Mode of irrigation

The mode of irrigation can affect not only the water content of the soil and thus the ion concentration of soil solution but also the leaching of ions from the soil. These will have an influence on the conductivity. From a comparison shown in Table 8–9 it is evident that the specific conductance of the cultivated layer in a plot with deep–water irrigation is the lowest, whereas a plot where the soil is kept wet is the highest. This trend in the difference between soils is in keeping with the difference in specific conductance of the percolating water, and is closely related to the water consumption of the field.

8.3.2.3 Plowing depth

Some examples in Table 8–10 show that under conditions where an equal amount of manure is applied, the specific conductance of the soil will be the lowest if the depth of plowing is the deepest due to the low amount of manure received in one unit weight of soil. Therefore, it can be frequently observed that if no increasing amount of manure is applied, deep-plowing can not result in an increase in yield. Rather, a drop in yield may even occur in many cases.

TABLE 8-9 Effect of mode of irrigation on electrical conductivity of the cultivated layer of paddy soil[1] (2-electrode method) (Changshu)

Mode of irrigation	Water consumption (mm)	Electrical conductivity of percolating water ($L \times 10^5$)	Electrical conductivity of soil ($L \times 10^5$)		
			Tillering stage	Elongating stage	Heading stage
Deep water-layer	2310	102	59.9	60.8	46.1
Shallow water-layer	2090	153	61.1	61.8	49.5
Intermittent	1390	—	75.8	77.8	69.2
Wet	1300	193	87.0	96.1	74.5

TABLE 8-10 Effect of deep cultivation and fertilization on electrical conductivity of paddy soils[1] (2-electrode method)

Locality	Treatment	$L \times 10^5$
Xiaogan	12 cm, lightly fertilized	38.0
	18 cm, lightly fertilized	27.3
	18 cm, moderately fertilized	34.6
	27 cm, moderately fertilized	30.0
	27 cm, heavily fertilized	35.7
Zixi	12 cm, lightly fertilized	21.4
	24 cm, lightly fertilized	15.0
	24 cm, moderately fertilized	22.2
	24 cm, heavily fertilized	34.4
	36 cm, heavily fertilized	18.7

8.3.3 Change in electrical conductivity due to the uptake of ions by rice

Results from numerous measurements show that while the electrical conductivity of the surface layer increases at the first stage of submergence, it decreases steadily after the transplanting of rice, and increases slightly again at the latter stage of rice growth. The causes responsible for this decrease in conductivity are twofold. The soluble ions may migrate downward along with the percolating

8.3 Electrical conductivity in relation to fertility

water. Furthermore the roots of rice absorb a large amount of nutrient. A special tank experiment (Fig. 8–11) in which the possibility of percolation was ruled out showed that for all the treatments the pattern of decrease in conductivity of a soil solution was in keeping with the decrease in the concentration of cations which consisted of 90—95% of calcium and magnesium, 2—10% of potassium, and less than 1% of ammonium and ferrous iron each. The slight increase in electrical conductivity at the latter stage of rice growth was due to the surplus of liberated nutrients over those absorbed by rice roots. From what has been mentioned above it can be suggested that for some paddy soils it should be possible to regulate the nutrient status of the soil through monitoring the electrical conductivity during the growing period of rice.

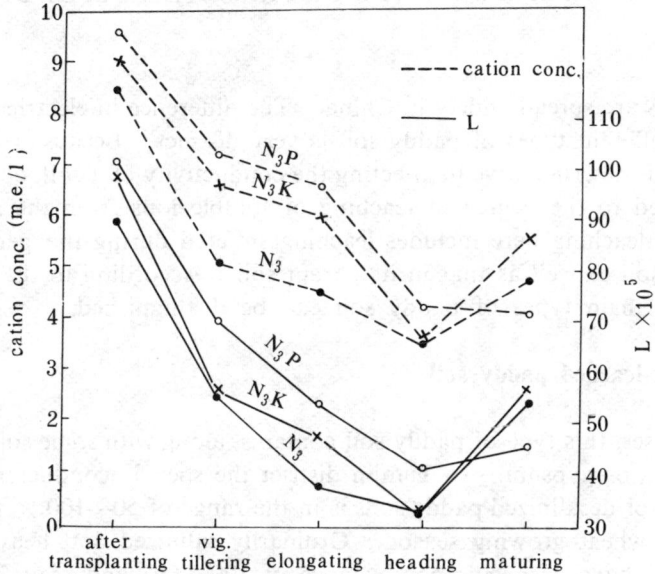

Fig. 8–11 Changes in electrical conductivity and cation concentration of soil solution during the growing period of rice[3]

8.3.4 Electrical conductivity and desalinization of salt-affected soils

As mentioned earlier, electrical conductivity is not capable of reflecting the fertility level of salt–affected paddy soils due to the masking of soluble salts such as sodium chloride over nutrient ions. However, we can consider the fertility problem of such soil from another viewpoint, namely toxicity. In salt–affected rice fields the concentration of soluble salts is frequently an important factor in determining the suitability for rice growth, and therefore, the degree of desalinization is an important index in evaluating the soil for agricultural use. The degree of desalinization of the soil reflects itself clearly in electrical conductivity. For example, in the coastal area of Tianjin district there is a rice field which has been well-desalinized through long-term rice plantation and has a specific conductance of less than 100×10^{-5} mho / cm within the whole profile. On the other hand, in a poorly-desalinized rice field the electrical conductance in the surface 0—20 cm layer exceeds 100×10^{-5} mho / cm, and in other horizons it exceeds 200×10^{-5} mho / cm. Intermediate between these two extremes is a field which has been desalinized in the upper layer, but not in the lower horizons. In this district, if the soil has a specific conductance of over 200×10^{-5} mho / cm, rice plants will usually show symptoms of salt toxicity. It has also been observed in the coastal area of Fujian Province that rice plants can not grow normally on soils with a specific conductance of over 200×10^{-5} mho / cm.

8.4 ELECTRICAL CONDUCTIVITY OF PRINCIPAL TYPES OF PADDY SOIL

Paddy soils are spread widely in China. The difference in electrical conductivity among different types of paddy soil is very distinct. Because the amount of soluble ions is determinative in affecting the conductivity of a soil, conductivity is closely related to the degree of leaching of soluble ions from the soil. The term degree of leaching here includes leaching suffered during the genetic process of paddy soil as well as that in its parent soil. According to the degree of leaching, three main types of paddy soil can be distinguished.

8.4.1 Weakly-leached paddy soil

In most cases, this type of paddy soil contains, along with some soluble salts, carbonates and / or gypsum. In Tianjin district the specific conductance of the cultivated layer of desalinized paddy soils is in the range of 50—100×10^{-5} mho / cm during the wheat-growing season. Ordinarily salinized and heavily salinized paddy soils have a specific conductance of 100—150 and over 200×10^{-5} mho / cm respectively. In the coastal area of Fujian Province the specific conductance of salt-affected paddy soils is also over 100×10^{-5} mho / cm, with an average value of 222×10^{-5} mho / cm for heavily salinized soils[4]. Fig. 8–12 shows the results for three representative profiles in Tianjin district. It can be seen that the electrical conductivity in the cultivated layer is rather high, although a large quantity of soluble salts has been leached to low-lying horizons due to long-term cultivation of rice. According to the statistics for 60 paddy fields in Sichuan and Hubei provinces developed on purplish soils, the mean value of the specific conductance is 70.2×10^{-5} mho / cm during the growing period of upland crops.

8.4.2 Moderately-leached paddy soil

In this type of soil the soluble salts and carbonates have been leached out, but the soil is essentially base-saturated, and has a pH of 6—7. According to the statistics based on 266 measurements made in the Suzhou district of Jiangsu

8.4 Electrical conductivity of principal types of paddy soil

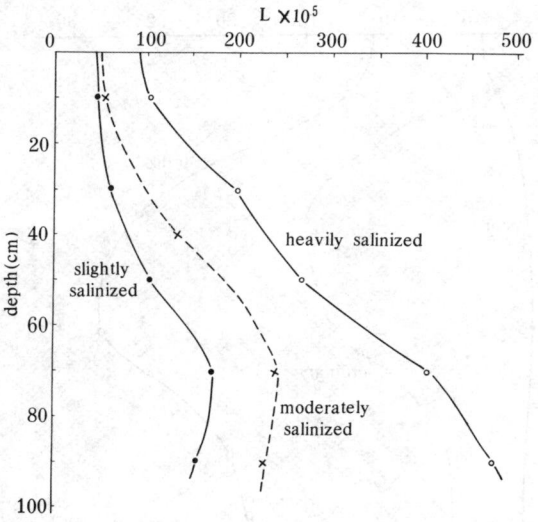

Fig. 8-12 Electrical conductivity in profiles of weakly-leached paddy soil (Tianjin, wheat-growing season)

Province, the mean value of specific conductance during the rice-growing season is 68.0×10^{-5} mho/cm. Fig. 8-13 shows the electrical conductivity of three paddy

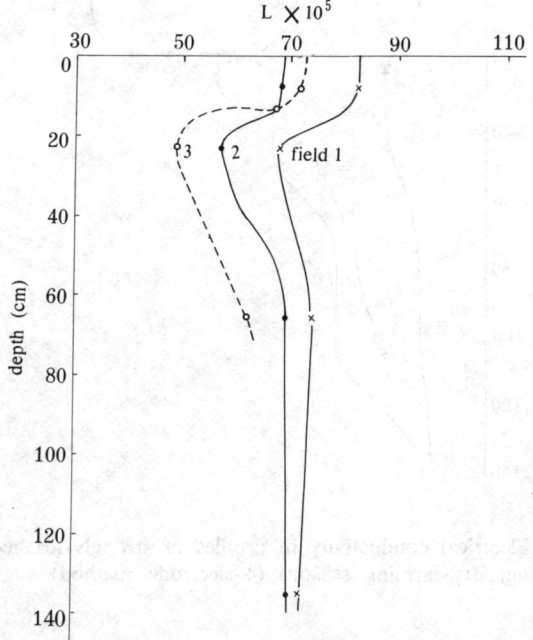

Fig. 8-13 Electrical conductivity in profiles of moderately-leached paddy soil (Jiangsu, rice-growing season) (4-electrode method)

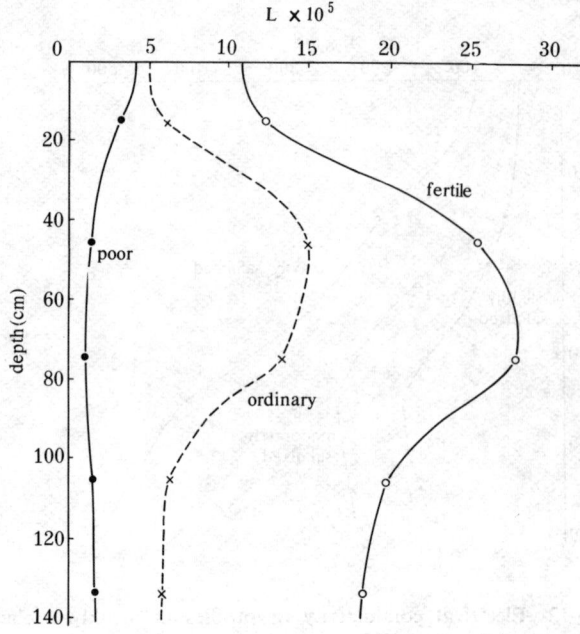

Fig. 8-14 Electrical conductivity in profiles of strongly-leached paddy soil (Jiangxi, rice-growing season) (4-electrode method)

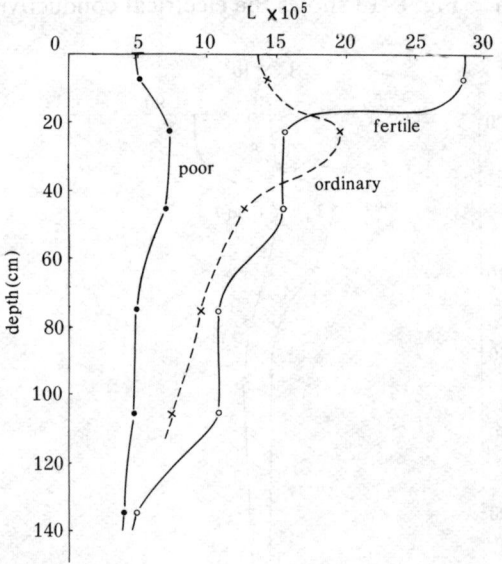

Fig. 8-15 Electrical conductivity in profiles of strongly-leached paddy soil (Zhejiang, dry-farming season) (4-electrode method)

profiles in Taihu region. The characteristic feature of this type of paddy soil is that the difference in conductivity among different fields is relatively small, and so is the case among different horizons within the profile. The electrical conductivity is slightly higher in the cultivated layer than in the plowpan, due apparently to the application of fertilizers.

8.4.3 Strongly-leached paddy soil

The base-saturation percentage of this type of soil is low, and so is the electrical conductivity. Taking paddy soils derived from red soils as an example, it was observed that the specific conductance in the cultivated layer of over 100 fields was mostly around 20×10^{-5} mho / cm during the rice-growing season[4], and in Taiho district of Jiangxi Province the mean value for 24 fields was 14.9×10^{-5} mho / cm. During the non-submerged season, the specific conductance of 7 paddy fields in Lanxi district of Zhejiang Province was below 10×10^{-5} mho / cm, and in Jinhua district of the same province only four out of the eight fields had a specific conductance of over 10×10^{-5} mho / cm. Paddy soils derived from yellow soils in Sichuan and Hubei provinces have a mean value of 31.6×10^{-5} mho / cm for 21 fields due to weak leaching as compared with those derived from red soils. Fig. 8-14 shows the electrical conductivity of several paddy profiles developed on red soils during the rice-growing season, while Fig. 8-15 pertains to the growing season of upland crops. It can be seen that the conductivity of this type of paddy soil is characterized by a low value in the cultivated layer, a large difference among various fields, and a wide difference among different horizons of the same profile. The pattern of the variation of conductivity within the profile may be distinguished into several types. Young soils are generally infertile, and the conductivity is low throughout the profile. In soils which have been cultivated with rice for some years, the electrical conductivity in the upper forty or fifty centimeters is relatively high, but in the lower horizons it is not much different from the parent red soil. In some profiles the conductivity is high in the middle part of the profile, but low in the cultivated layer and the lower part of the profile. For well-developed profiles the electrical conductivity in the middle and lower parts is high, but remains lower in the cultivated layer. However, in fertile soils which have been heavily fertilized the conductivity in the cultivated layer may be quite high. These patterns of the variation in conductivity within the profile are in reality a reflection of the variation in base-saturation percentage of the soil (cf. Chapters 6 and 9).

REFERENCES

(1) Institute of Soil Science, 1961. Soil Environment of High-yield Rice. Chapters 9,11,12. Science Press, Beijing.
(2) Yu Tian-ren et al., 1976. Electrochemical Properties of Soils and Their Research Methods (revised ed.). Chapter 7. Science Press, Beijing.
(3) Yu Tian-ren et al., 1959. Studies on electrochemical properties of soils. I. Electrical conductivity of paddy soils and its relation to soil fertility. Acta Pedologica, 7: 145-158.
(4) Wu Deng-sheng and Guan Wen-fen, 1979. Changes in electrical conductivity of the sap of rice. Soils, 3: 19-21.

CHAPTER 9

PHYSICAL CHEMISTRY OF PADDY SOIL IN RELATION TO SOIL GENESIS

Yu Tian-ren

Paddy soil is the result of a series of physical, chemical and biological processes taking place in various parent (preceding) soils after the cultivation of rice. Paddy soil inherits some properties from its parent soil, and is also affected by pedogenetic processes of the paddy soil itself. And, the better-developed the paddy soil, the more pronounced this effect.

The pedogenesis of paddy soil is closely related to water. The alternation of wetting and drying during the year induces periodical changes in the oxidation-reduction regime and a series of other physico-chemical properties of the soil. All these changes exert their effects on the genesis of paddy soil.

In this chapter, some physico-chemical processes which can affect the genesis of paddy soil are analysed first, then the change in the composition of the solid part of the soil —— clay minerals induced by these processes is discussed, and, finally, based on the principles of soil classification deduced from theories discussed in this book, principal types of paddy soil of China are presented.

9.1 GENESIS OF PADDY SOIL

From the chemical point of view, the pedogenetic process of paddy soil is chiefly a process of the leaching of materials. According to their degree of activity and relative importance to the soil, the physico-chemical mechanisms involved in the leaching processes may be analysed as solution, reduction and complexation, which will be discussed in order.

9.1.1 Solution

The quantity of water coming into contact with each unit weight of paddy soil is much more than that for upland soils. For instance, if an upland soil is assumed to contain 18% water and a paddy soil 40%, the latter would more than double the former. During the submerging period, the water will percolate downward under the influence of gravity. In paddy fields the total amount of percolation water is generally over one thousand millimeters per year, differing from upland soils in which only a small amount of water is percolated during the rainy season. Besides, the high temperature in the submerging season is also favorable for solution reactions. All these factors help make the solution of

9.1 Genesis of paddy soil

elements active in paddy soils.

The solubilities (including the distribution between the dissociated and adsorbed forms) of various elements in the soil are quite different. Their contents also differ to a large extent. As a consequence, the concentration of various ions in soil solution varies greatly. Table 9–1 shows the chemical composition of some percolation waters in an acid paddy soil region. It can be seen that the concentrations of aluminum and phosphate are very low, and that soluble silicon accounts for only a very small proportion of the total silicon of the soil. The most important elements leached out with percolation water are calcium, magnesium, potassium and sodium. Since the total content of potassium in these soils is quite low, it is interesting to note that the intensity of the leaching away of potassium is very high, although the absolute amount of leached potassium is not so high as that of calcium or magnesium. The concentration of water–soluble potassium increases drastically after fertilization (Table 9–1). This rapid leaching loss of fertilizer–potassium can also be observed in the dynamic change of exchangeable potassium in the soil profile after fertilization. Fig. 9–1 shows the results in this respect for a paddy soil profile in a mountain region of Jiangxi Province. At the tillering stage of rice growth, some of the potassium applied in the cultivated layer have been accumulated in the plowpan, and at the booting stage they have been accumulated in lower horizons. At the same time, the content of exchangeable potassium in the cultivated layer decreases remarkably. The leaching loss of calcium and magnesium is also quite conspicuous, although not so marked as in the case of potassium. In Fig. 9–2 is shown the distribution of exchangeable calcium in an acid paddy soil derived from red soil after the application of different amounts of lime. It is seen that after liming for five years the accumulation of calcium is evident at a depth of 50 cm or even 70 cm.

TABLE 9–1 Chemical composition of percolating waters[2] (Jinxian)

Water	Composition (m mole / l)						
	Ca	Mg	K	Na	Al	SiO_2	P
Irrigation water	0.28	0.053	0.038	0.026	—	0.025	0.0001
Percolating water of red soil	0.11	0.049	0.015	0.22	tr.	0.27	0.0008
Percolating water of paddy soil (before fertilization)	1.48	0.79	0.079	0.17	0.01	0.21	0.015
Percolating water of paddy soil (after fertilization)	2.87	1.05	2.37	0.24	—	0.20	0.02

Thus, it is clear that for salt–affected paddy soils the most distinct chemical change in the initial stage of soil development is the leaching away of chloride, sodium and other water–soluble ions, and for paddy soils derived from strongly acid soils containing only a small amount of exchangeable bases, such as red soil, the most distinct chemical change in the early stage of soil development is

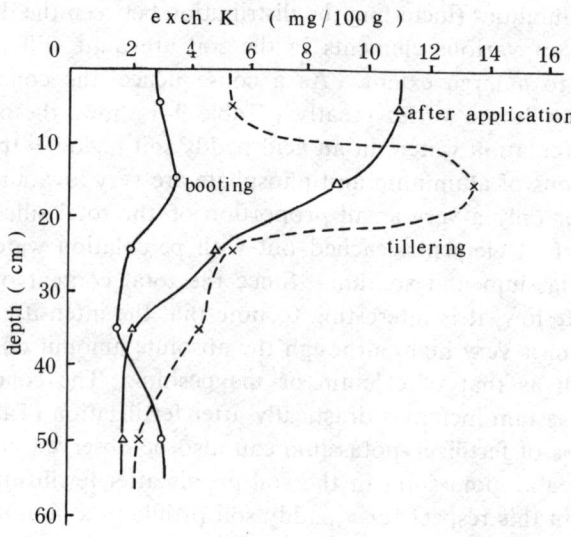

Fig. 9-1 Dynamics of exchangeable potassium in a paddy soil profile (late rice, Jiangxi)[1]

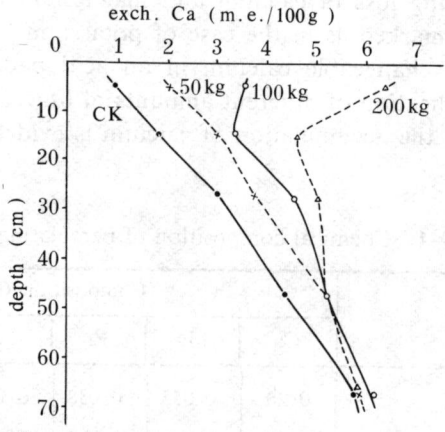

Fig. 9-2 Exchangeable calcium in an acid paddy soil profile five years after the application of different amounts of lime (courtesy of M. T. Ma and J. C. Xie)

the redistribution of bases coming from fertilizers in the profile. Fig. 9-3 shows such an example. The base-saturation percentage of the red soil is only 10—20%. At the early stage of soil development, the bases accumulate only in the plowpan. At its middle stage a large amount of bases is accumulated in the middle part of the profile, and at its later stage the accumulation extends to the lower horizons. The three patterns of the distribution of bases in the profile shown in Fig. 9-3,

9.1 Genesis of paddy soil

together with another pattern in which the base-saturation percentage is high throughout the profile, are the main patterns representative of paddy soils of South China. The main cause of the redistribution of basic cations in the profile is the solution of these ions.

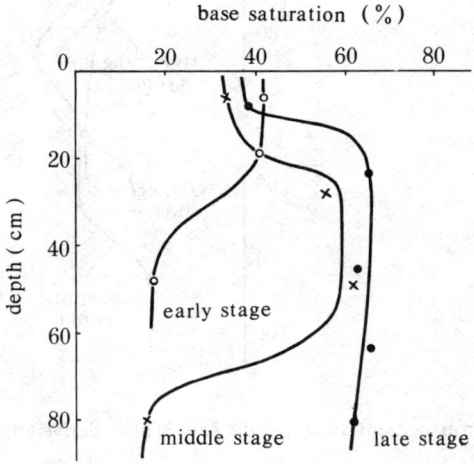

Fig. 9-3 Base-saturation percentage of paddy soil at different stages of soil development[6]

9.1.2 Reduction

Some elements are practically inactive when existing in the oxidized form owing to their low solubility in water. However, when reduced, the mobility may increase strikingly, and will play an important role in soil genesis. As far as paddy soil is concerned the most important elements in this respect are iron and manganese. For these elements, the concentrations of their high-valent cations such as Fe^{3+} and Mn^{4+} in soil solution are so low that they can be ignored even for strongly acid soils with a pH of 4—4.5. Under reduced conditions, the amounts of ferrous and manganous ions are so high that they can in some cases exceed the amount of exchangeable bases. The physico-chemical properties of iron and manganese have been discussed in detail in Chapter 4. In this section, some questions relating to soil genesis will be considered.

It has been shown in Chapter 4 that manganic manganese is reduced more easily than ferric iron, and that under submerged conditions most of the manganese is reduced to the manganous state. The manifestation of this difference in the chemical behavior of these elements in soil genesis is that the eluviation and illuviation of manganese are more distinct than those of iron. This can be seen from distributions of iron and manganese in three paddy profiles shown in Figs. 9-4 and 9-5. For a young paddy soil which has developed for four years on a red soil (Hengyang), if the content of iron or manganese in the cultivated layer is taken as 100%, then the relative content of iron in the illuvial

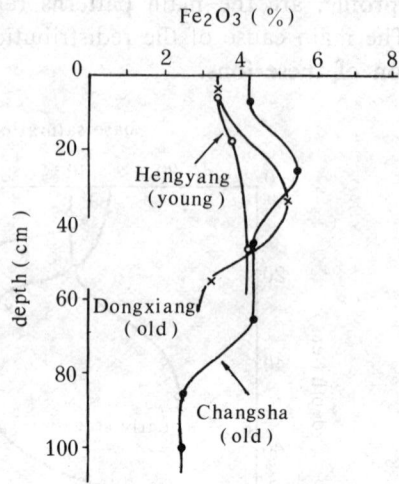

Fig. 9-4 Iron content of a paddy soil profile derived from red soil[6]

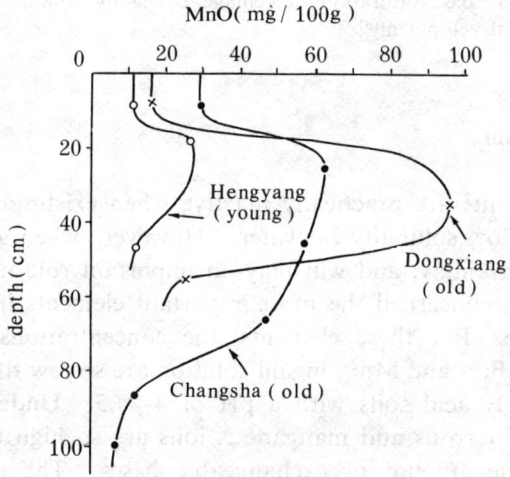

Fig. 9-5 Manganese content of a paddy soil profile derived from red soil[6]

horizon is 110%, and that of manganese is 251%. For a well-developed paddy soil (Dongxiang), the relative contents of iron and manganese in the illuvial horizon referring to the cultivated layer are 156% and 666% respectively. As a consequence, the iron to manganese ratio in concretions which are the result of the accumulation of iron and manganese compounds is much smaller than that in the original soil (Table 9-2).

9.1 Genesis of paddy soil

TABLE 9-2 Comparison of iron/manganese molecular ratio between Fe–Mn concretion and ordinary soil[3]*

Locality	Specimen	Content (%)		Fe_2O_3 / MnO
		Fe_2O_3	MnO	
Jiangxi	Concretion	7.22	0.50	6.35
Jiangxi	Concretion	15.90	1.47	4.80
Jiangxi	Hardpan	9.31	1.07	3.85
Jiangxi	Hardpan	18.32	5.68	1.43
Zhejiang	Concretion	13.39	2.50	2.37
Jiangsu	Concretion	12.73	1.75	3.22
Jiangsu	Concretion	10.4	6.00	0.77
Ordinary soil**		6	0.15	17.4

* Some data from Su Qi
** Assumed value

The data presented above mean that the differentiation of iron and manganese in paddy profiles is conspicuous, especially for well–developed soils. Because this differentiation of iron and manganese in the profile is a distinct morphological and chemical index of the development of paddy soil, it is possible to judge the stage of development of a paddy soil from the distribution of iron and manganese in the profile.

The significance of eluviation–illuviation of iron and manganese in soil genesis is not confined to the differentiation of these elements itself. This differentiation may have obvious influences on morphological features of the soil such as color, structure and consistency. Hence in the designation of pedogenetic horizons within the profile the content and form of iron and manganese are frequently used as an important and, in some cases, the principal index. The eluviation or illuviation of iron and manganese is dependent on the oxidation–reduction status of the given horizon, and the latter is in turn determined chiefly by the water regime. The water of a paddy soil comes through irrigation or from the ground water. As ground–water is ordinarily affected by irrigation water to some extent, it is also related to the planting of rice. Irrigation water induces reduction processes in the surface layer, and ground water causes the same in the lower horizons. Therefore, from the viewpoint of the eluviation–illuviation of iron and manganese, four basic pedogenetic horizons are identified in a paddy soil profile, namely, Ag (abbreviated as A), B, G, and C horizons. A–horizon is the cultivated layer, and is an eluviation horizon with respect to iron and manganese. G–horizon is also an eluviation horizon caused by the movement of ground water. C–horizon still preserves the basic characteristics of the parent soil. In the upper part of the B–horizon connecting with the A–horizon, the soil is relatively compact owing to the effect of mechanical pressure by plowing, and this part has a special name of plowpan (P) horizon. The thickness of this horizon is generally 10—20 cm. As the history of the P–horizon is relatively short when considered as an illuvial horizon, the accumulation of iron and manganese in this horizon is not so pronounced as that in the main solum of the B–horizon. In some paddy profiles the reduction condition develops to a certain degree in the P–horizon at the later period of submergence, but generally speaking the illuviation process of iron and manganese prevails in this horizon. In this horizon there is also accumulation of bases. The rest of the B–horizon besides P–horizon may be further distinguished as subhorizons B_1 and B_2. The color of the subhorizon B_1 is usually grayer than that of the subhorizon B_2 owing to the presence of some gray coatings formed by some substances leached down from the cultivated layer, and to the weakening of the oxidation condition at the later period of rice growth.

In the upper part of the G–horizon there may sometimes exist a BG–horizon. This horizon is deeply affected by ground water, but at the dry season the lowering of the water table may also induce the appearance of a weak oxidation condition in this horizon. As a result, the morphological feature of this horizon is that there are mottlings of gray color as well as of brown–black color. It can be seen from the analytical data of Table 9–3 that the content of iron in this horizon is lower than that in the parent soil, although slightly higher than that in the G–horizon, and that the content of manganese is higher than that in the parent soil, especially in the G–horizon. It may be assumed that the accumulated manganese in this horizon comes mainly from the G–horizon dissolved in ground water.

TABLE 9–3 Contents of iron and manganese in BG, G and C horizons of paddy soil[5]

Locality	Horizon	Fe_2O_3 (%)	MnO (mg / 100g)
Changsha	BG	4.39	94
	G	2.53	13
	C	5.65	50
Dongxiang	BG	3.15	52
	C	7.13	41
Yiwu	BG	3.87	103
	G	2.67	46
	C	5.54	62

9.1.3 Complexation

It has been known from Chapter 4 that iron and manganese can complex

9.1 Genesis of paddy soil

(including chelate) with certain organic substances in the soil. The significance of this complexation in the genesis of paddy soil should not be overlooked. From the viewpoint of the eluviation of iron and manganese, this complexation differs from the reduction process in that the complexation reaction itself does not lead to the change in the valent state of iron or manganese, but it can promote the transfer of these elements from the solid phase to the solution owing to the strong affinity of the organic complexing agents for iron and manganese. Besides, the formation of complexes of reduced iron and manganese leads to an increase in the concentration of these elements in solution, and hence complexation is frequently a reaction superimposed on reduction, thus helping the solution of iron and manganese.

Table 9–4 shows the experimental results of the complexation of the decomposition products of two plant materials with iron of the soil. After the elimination of organic reducing substances by aeration, the remaining part could still cause a considerable amount of iron to transfer to solution through its interaction with the soil for several hours, especially for the decomposition products of vetch. When comparisons are made among the three soils it can be seen that the amount of complexed iron is the largest for the paddy soil derived from lacustrine deposits of the Taihu region containing a higher percentage of iron oxides. From this experiment it must be admitted that complexation may play an important role in the pedogenetic process of paddy soil.

TABLE 9–4 Complexation of iron of paddy soil by decomposition products of organic matter (data from Wu Shi–han)

Parent soil	Complexed iron (m.e. / 100g)	
	Rice straw*	Vetch**
Yellow soil	0.013	0.01
Red soil	0.027	0.05
Lake deposit	0.032	0.15

* Eh 550 mV, pH 5.6
** Eh 520 mV, pH 5.5

The ease with which iron and manganese go into complexation in different horizons is related to the form of the oxides of these elements. From the experimental result of Table 9–5 it can be seen that the iron of the cultivated layer is easier to be complexed than that of the illuvial horizon. This is perhaps due to the lower degree of "ageing" of iron oxides in the former case. The iron of the parent red soil is very difficult to complex by EDTA owing to its high degree of ageing. It is for the same reason that the amount of manganese capable of being complexed in the red soil is very small. The low content of complexable manganese in the cultivated layer is due to the fact that most of the active manganese has been lost through eluviation.

TABLE 9-5 Complexing capacity of iron and manganese in different horizons of paddy soil with EDTA[5]

Horizon	Iron		Manganese	
	mg/100g	In total (%)	mg/100g	In total (%)
A	22.3	0.93	3.6	14.2
B	13.6	0.37	51.8	31.3
BG	9.6	0.43	21.3	53.1
C	7.5	0.14	2.8	8.6

As to the significance of complexation in the eluviation of calcium and magnesium in the soil, it is not yet clear.

9.2 CHANGE IN CLAY MINERALS DURING THE GENESIS OF PADDY SOIL

Under the long-term influence of the pedogenetic process of paddy soil, the clay minerals suffer changes to varying degrees. These changes are the combined results of the destructive action of solution, reduction and complexation, the recombination of some elements, and the leaching of some dispersed soil clays. Among these, the process of depotassication of potassium-bearing minerals and the process of deferrugination of iron-rich minerals are perhaps the most distinct. Other soil minerals may also undergo some changes. These will be discussed in the following section.

9.2.1 Depotassication

Depotassication is a result of the long-term dissolution of potassium-bearing minerals. In paddy soils three cases may be distinguished.

(a) In soils rich in potassium-bearing minerals, the change in clay minerals during the genesis of paddy soil is conspicuous owing to heavy depotassication. Taking a paddy soil derived from purplish soil in which illite is predominant with a small amount of vermiculite and kaolinite as an example (Table 9-6),

TABLE 9-6 Properties of the clay fraction of a paddy soil derived from purplish soil (Jiangxi)[10]

Sample	K_2O (%)	C.E.C. (m.e./100g)	Combined water (%)	Dominant clay minerals
Parent soil	5.25	13.2	7.3	Ill. with some kl. and verm.
Surface layer	3.74	18.8	9.9	Ill. less and kl. and verm. more than parent soil
Substratum	3.64	18.3	9.2	Ill. less and kl. and verm. more than parent soil

9.2 Change in clay minerals during the genesis of paddy soil

the content of illite decreases, while that of vermiculite and kaolinite increases as compared with the parent soil. The K_2O content of the clay fraction decreases by about 1.5%, and the cation exchange capacity increases by about 5 m.e. / 100g.

(b) In soils with a medium amount of potassium-bearing minerals, the process of depotassication is not so distinct as that in case (a). For example, in a paddy soil derived from neutral loess (yellow-brown soil), the K_2O content is lower by 0.5% than that in the parent soil, while the cation-exchange capacity is higher by 2—3 m.e. / 100g (Table 9-7). The clay minerals are mainly illite and montmorillonite along with small amounts of vermicullite and kaolinite in both the paddy soil and the parent soil. However, in the paddy soil the content of illite is slightly lower.

(c) In highly-weathered soils with few potassium-bearing minerals, the content of potassium and cation exchange capacity of the clay fraction are higher in the paddy soil than in the parent soil (Table 9-8).

TABLE 9-7 Properties of the clay fraction of a paddy soil derived from yellow-brown soil (Jiangsu)[10]

Sample	K_2O (%)	C.E.C. (m.e. / 100g)	Combined water (%)	Dominant clay minerals
Parent soil	3.06	39.3	10.6	Ill. and mt. with some kl. and verm.
Surface layer	2.67	42.7	9.2	Ill. slightly less than parent soil
Substratum	2.54	41.0	10.2	Ill. slightly less than parent soil

TABLE 9-8 Properties of the clay fraction of paddy soils derived from lateritic soils (Guangdong)[8]

Parent soil	Sample	K_2O(%)	C.E.C. (m.e./100g)	Combined water (%)	Clay minerals
Lateritic soil	Parent soil	1.04	7.2	16.4	Kl. predominant, with some ill. and gib.
	Surface layer	1.11	12.9	—	Gib. less than parent soil
	Substratum	1.26	11.5	13.8	Gib. less than parent soil
Laterite	Parent soil	0.10	5.2	16.1	Kl. and gib.
	Surface of young paddy soil	0.38	10.9	—	As parent soil
	Substratum of young paddy soil	1.29	9.6	17.5	As parent soil
	Surface of old paddy soil	0.17	24.6	13.7	Gib. and kl. less than parent soil, mt. present
	Substratum of old paddy soil	0.15	20.4	14.8	As above
	Gley horizon of old paddy soil	0.17	17.9	15.5	As above

It is not yet known whether the increase in the potassium content of the clay fraction is due to a process of "re-potassication" of clay minerals in the presence of soluble potassium coming from fertilizers and irrigation water, or due to the potassium-bearing minerals being brought into the soil, or to both.

9.2.2 Deferrugination

The process of deferrugination is caused chiefly by reductive eluviation. Complexation can also promote the advancement of this process. Because this process will result in a decrease of the amount of iron-bearing minerals in the clay fraction, it is generally observed that in the cultivated layer of paddy soil which has suffered reductive eluviation the iron content is lower than that in the illuvial horizon or parent soil. Table 9-9 shows some comparisons in this respect.

It is possible to use the Al_2O_3/Fe_2O_3 molecular ratio of the clay fraction as an index for expressing the degree of deferrugination. As seen from Table 9-9, the Al_2O_3/Fe_2O_3 ratio in the cultivated layer is much higher than that in the illuvial horizon or the parent soil. The clay minerals of the glei horizon have also suffered the process of deferrugination. Similarly, the plinthitic horizon with intermingled white- and red-colored mottlings in the lower part of some red soils is also a product of gleization. It can be seen from comparisons shown in Table 9-10 that in the glei horizon of paddy soils and the white-colored fraction of plinthite, the Al_2O_3/Fe_2O_3 ratio is much higher than that in the parent soil or the red-colored fraction of plinthite.

TABLE 9-9 Comparison of chemical composition between surface layer and illuvial horizon (or parent soil) of paddy soils[2, 3]

Parent soil	Locality	Horizon	$Fe_2O_3(\%)$	Molecular ratio		
				SiO_2/Al_2O_3	SiO_2/Fe_2O_3	Al_2O_3/Fe_2O_3
Lateritic soil	Guangxi	Surface	1.69	1.81	79.8	44.1
		Illuvial	14.1	1.88	8.51	4.53
Red soil	Guangxi	Surface	6.05	2.18	12.10	5.55
		Parent soil	8.49	2.19	10.4	4.77
Red soil	Taiwan	Surface	10.4	2.73	12.1	4.40
		Illuvial	20.0	2.64	5.38	2.04
Red soil	Jiangxi	Surface	7.89	2.21	14.7	6.64
		Parent soil	12.2	2.15	8.94	4.16
Yellow-brown soil	Sichuan	Surface	8.40	3.06	15.5	5.05
		Illuvial	19.4	2.55	5.44	2.13

TABLE 9-10 Chemical composition of the glei horizon of paddy soil and the plinthitic horizon of red soil[10]

Parent material	Locality	Sample	Fe_2O_3(%)	Molecular ratio		
				SiO_2/Al_2O_3	SiO_2/Fe_2O_3	Al_2O_3/Fe_2O_3
Laterite	Guangdong	Gley	9.26	2.11	11.8	5.59
		Parent soil	18.14	1.55	4.44	2.87
Red soil	Jiangxi	Gley	2.97	2.14	41.3	19.3
		Parent soil	12.74	2.22	8.55	3.85
Marine deposit	Guangdong	White fraction	3.92	2.27	31.9	14.1
		Red fraction	9.17	2.38	13.2	5.55
Metamorphic rock	Guangdong	White fraction	4.77	2.46	25.9	10.5
		Red fraction	14.42	2.34	7.5	3.21
Red soil	Jiangxi	White fraction	4.06	2.42	31.8	13.1
		Red fraction	19.77	2.39	5.3	2.22

The degree of deferrugination in the clay fraction of paddy soils increases with the advancement of soil development. Fig. 9-6 shows that in the clay fraction of the cultivated layer the iron content is lower to a greater extent than that in the parent soil for soils with heavier deferrugination. On the other hand, in the illuvial horizon (including plowpan), the iron content is higher to a greater extent for soils which suffered heavier deferrugination.

The change in composition of the clay fraction of paddy soils caused by deferrugination has an important influence on the electric charge and a series of other related physico-chemical properties.

The behavior of manganese is similar to that of iron. However, it is not so important as iron in paddy soils owing to its lower content.

9.2.3 Other changes

From the viewpoint of mechanical composition, solution, reduction and complexation all have a destructive action on soil clay. The periodical change

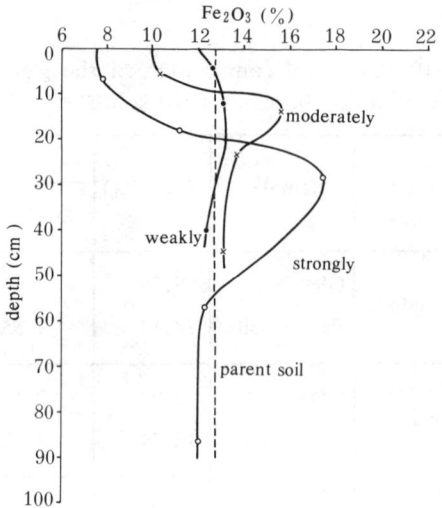

Fig. 9-6 Iron content of the clay fraction of paddy soils with different degrees of deferrugination[9]

in pH caused by the alternation of oxidation and reduction conditions of the soil can also lead to the destruction of clay minerals. Taking the cultivated layer of an acid paddy soil derived from red soil as an example, if it is assumed that the pH during the submerged season is 6.5 and returns to 5 again during the drained season, and assuming that the cation-exchange capacity of the soil is 8 m.e./100g, it can be estimated that there would be about 4 m.e./100g of aluminum liberated in a cycling of oxidation-reduction owing to the corrosion of clay by hydrogen ions[4] which might appear during the reacidification of the soil. Computing in terms of aluminum oxide, this corresponds to 68 mg, accounting for 0.068% of the clay by weight. Of course, owing to the reprecipitation of aluminum caused by the rise in pH after submerging, the chemical leaching of aluminum under actual field conditions is not strong (cf. Table 9-1). Nevertheless, the repetition of the acidification-neutralization circle of the soil during the year will undoubtedly result in a decrease in the content of clay. Besides, the mechanical leaching of soil colloid under the influence of moving water should be noticeable. Therefore, as a rule, in the cultivated layer or the glei horizon which has been affected by water to a greater degree the clay content is always lower than that in the parent soil, as is evidenced by comparisons shown in Table 9-11.

It can be observed in Tables 9-6, 9-7 and 9-8 that the cation-exchange capacity of the clay fraction of paddy soil is higher than that of the respective parent soils. Similarly, the cation-exchange capacity of the white-colored fraction of the plinthite is higher than that of the red-colored fraction (Table 9-12). These seem to imply that the reductive eluviation can cause the increase in the cation-exchange capacity of the clay. Since it is seen from Tables 9-9 and 9-10 that for paddy soils derived from red soils the silica/alumina ratio differs insigni-

9.2 Change in clay minerals during the genesis of paddy soil

TABLE 9-11 Mechanical composition of different horizons of paddy soil[6]

Locality	Horizon	Mechanical composition (%)		
		2—0.02 mm	0.02—0.002 mm	<0.002 mm
Changsha	A	18.0	52.0	30.0
	G	24.5	52.5	23.0
	C	8.5	40.0	51.5
Yiwu	A	17.0	57.0	26.0
	G	13.0	67.0	20.0
	C	12.0	55.0	33.0
Hengyang	A	21.0	46.0	33.0
	C	6.0	30.0	64.0

TABLE 9-12 Cation–exchange capacity of different color fractions of plinthite (clay fraction)

Parent material	C.E.C. (m.e./100g)	
	Red fraction	White fraction
Marine deposit	15.8	17.6
Metamorphic rock	9.8	13.0
Red clay	18.1	26.7

ficantly among various horizons of the same profile and between the paddy soil and its parent soil, and since it is also difficult to demonstrate the differences in the composition of crystalline clay minerals by differential thermal analysis or X-ray diffraction analysis, there arises the question of the real cause of the increase in the cation–exchange capacity of paddy soil after undergoing the reduction reaction. It is not yet known whether it is caused by a real change in the compositon of crystalline clay minerals, or by an increase in effective negative charge induced by the loss of ferric oxides, or by both actions. It should be remembered that the above discussions are confined to the clay fraction of the soil. As far as the whole soil is concerned, the cation–exchange capacity of paddy soil is generally lower than that of the parent soil, if the influence of organic matter is excluded.

9.3 CLASSIFICATION AND TYPES OF PADDY SOIL

9.3.1 Principles of classification

Because paddy soils are developed on different parent soils and have undergone various pedogenetic processes in different degrees, there are a variety of types of paddy soil in nature. These soils should be classified chiefly on the basis of current properties of the soil. The properties of paddy soil partly originate in the parent soil and partly result from the pedogenetic processes. These two factors should be considered comprehensively in the classification. From the viewpoint of either the genesis of the parent soil or that of the paddy soil, eluviation processes play the dominant role. Therefore, in the classification of paddy soils the degree of eluviation should be considered as the principal criterion.

Although different in mechanism, solution eluviation and reductive eluviation are interrelated. For parent soils not affected by submergence, the solution eluviation plays a key role. In the genesis of paddy soil, the two processes proceed at the same time. However, solution eluviation is generally more active, and may exert influences on reductive eluviation. This is because the solubility of ferrous and manganous ions formed under reducing conditions is strongly dependent on pH, and the pH of a soil is in turn determined chiefly by the degree of the leaching away of bases. Therefore, it may be observed that reductive eluviation is more distinct in paddy soils which have been subjected to strong solution eluviation. It should also be remembered that for the majority of paddy soils the eluviation strength undergone in the course of genesis is consistent with that of the parent soils. Considering these two factors together, it may be concluded that in the classification of paddy soils solution eluviation should be used as the criterion for the higher category of classification, and reductive eluviation for the lower category of classification.

The consequences of solution eluviation find expression chiefly in the contents of basic elements such as calcium, magnesium, potassium and sodium, including both the insoluble and the ionic (exchangeable and water-soluble) parts. Based on this index of solution eluviation (leaching), paddy soils may be classified into three main types, namely weakly-leached, moderately-leached and strongly-leached.

Within each main type, the soils may be further divided into three subtypes, namely oxidizing, redoxing and reducing, according to the oxidation-reduction regime of the soil. In oxidizing paddy soils, except during the growing period of rice when the soil is submerged, the oxidation-reduction condition within the one-meter solum is in an oxidized state, and the profile has a structure of APB or APBC. In redoxing soils, there appears a glei horizon beneath the illuvial horizon, and the profile has a structure of APBG. In reducing paddy soils, the glei horizon lies near the plowpan or the cultivated layer, and the profile structure is of the APG or AG type.

It is possible to further distinguish soil families within each subtype according to other properties of the soil.

9.3.2 Weakly-leached paddy soil

The weakly-leached type of paddy soil is derived from salinized or calcareous parent soils. Because of the short history of rice cultivation and the short duration of submergence in a year, and because the low temperature is unfavorable to weathering processes, the soil contains large amounts of basic elements in the form of minerals, together with some calcium carbonate and/or soluble salts. The CaO content of the soil is generally over 2%, with about 2% of K_2O and more than 1.5% of Na_2O. The soil is base-saturated, with a pH of over 7.5 when not submerged. The eluviation-illuviation of iron and manganese in this type of soil is not distinct. In Table 9–13 are shown the analytical results for such a profile developed on recent alluvial material from the Yangtze River containing carbonate.

TABLE 9–13 Chemical composition of a weakly-leached paddy soil (Shazhou)*

Depth (cm)	Horizon	Composition (%)				
		CaO	MgO	K_2O	Na_2O	P_2O_5
0—20	A	3.71	2.16	2.19	1.40	0.176
20—28	P	3.82	2.20	2.17	1.39	0.160
28—54	B	3.54	2.35	2.18	1.35	0.144
54—70	C	4.10	2.10	2.12	1.47	0.141

* Data from Su Qi

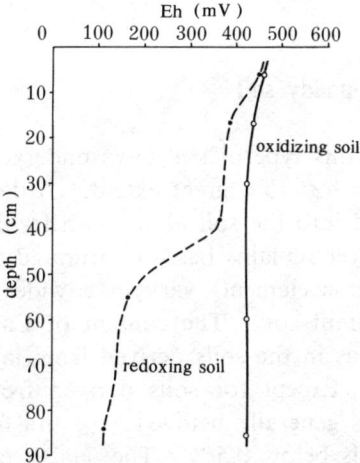

Fig. 9–7 Eh of weakly-leached paddy soil profiles with different oxidation-reduction status (Beijing)[2]

Within this type of soil three subtypes may be identified according to their oxidation–reduction conditions. Fig. 9-7 shows the variations of oxidation–reduction potentials of two paddy profiles in the Beijing region.

9.3.3 Moderately-leached paddy soil

In the moderately–leached type of paddy soil the soluble salts and calcium carbonate have been leached out. The contents of CaO, K_2O and Na_2O are 0.8—2%, 1.5—2% and about 1% respectively. In the profile there is differentiation of iron and manganese. but the differentiation is not so distinct as in the strongly–leached paddy soil. Table 9-14 shows the analytical results of a representative profile of such a type.

The oxidation–reduction potentials of the three subtypes of such a paddy soil have already been shown in Fig. 1-13.

TABLE 9-14 Chemical properties of a moderately–leached paddy soil (Wuxian)

Depth (cm)	Horizon	pH	Composition* (%)				Exchangeable bases** (m.e./100g)			
			CaO	MgO	K_2O	P_2O_5	Ca	Mg	K	Na
0—15	A	6.0	0.82	1.00	1.71	0.25	13.4	4.3	0.21	0.32
15—23	P	7.0	0.97	0.95	1.84	0.23	15.1	4.3	0.16	0.26
23—40	B_1	7.0	0.99	1.01	1.73	0.14	14.7	5.1	0.17	0.26
40—60	B_2	7.1	0.96	1.06	1.76	0.13	14.2	3.5	0.15	0.23
60—80	Cg	7.0	0.87	1.00	1.48	0.05	15.2	5.4	0.19	0.31

* Data from Su Qi
** Data from Zhang Kuo–zhu

9.3.4 Strongly-leached paddy soil

The parent soils of this type of soil have undergone strong eluviation, and basic elements have been lost to a great extent. Although a certain amount of bases may be introduced into the soil along with fertilizers after the cultivation of rice, the cultivated layer remains base–unsaturated and has a pH of less than 6.5. The contents of base elements vary in a wide range, depending chiefly on the nature of the parent soil. The content of CaO is generally below 1%, and in some cases such as in the soils derived from laterite or lateritic soil it is only in trace amounts. Except for soils derived from potassium–rich parent soils, the K_2O content is generally below 1.5%, and for those derived from laterite or lateritic soil, it is below 0.5%. The Na_2O content is generally below 1%, and for those derived from red soils it is always below 0.5%. The eluviation and illuviation of iron and manganese in this type of soil is quite distinct. In Table 9-15 are shown the analytical results for such a representative profile.

9.3 Classification and types of paddy soil

TABLE 9-15 Chemical properties of a strongly-leached paddy soil (Taoyuan)*

Depth (cm)	Horizon	pH	Composition (%)					Exchangeable bases (m.e./100g)			
			CaO	MgO	K_2O	Na_2O	P_2O_5	Ca	Mg	K	Na
0—12	A	5.4	0.68	0.67	1.14	0.36	0.10	8.75	0.93	0.16	0.15
20—35	P	6.0	0.66	0.77	1.14	0.28	0.05	8.13	1.15	0.13	0.13
45—60	Bg	6.8	0.64	0.74	1.25	0.35	0.04	7.25	1.20	0.09	0.12
70—85	B	7.0	0.84	0.69	1.24	0.42	0.02	7.13	1.25	0.12	0.11

* Data from Gong Zi-tong

Examples of the oxidation-reduction potentials of three subtypes of strongly-leached paddy soil can be found in Fig. 9-8.

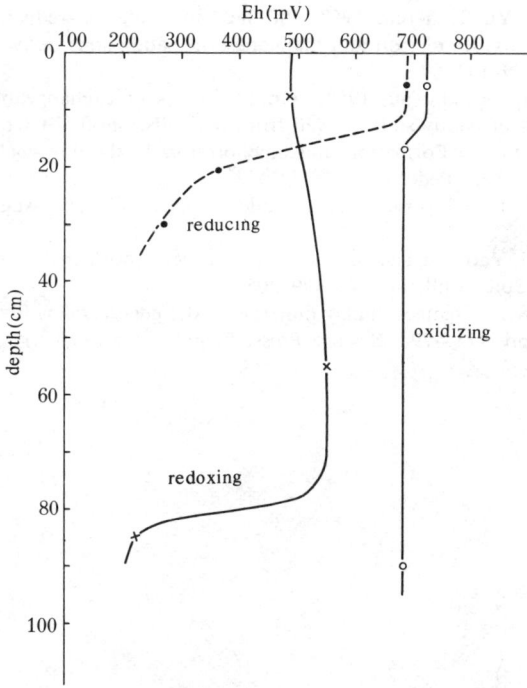

Fig. 9-8 Eh of strongly-leached paddy soil profiles with different oxidation-reduction status (Guangdong)

Among the three types of paddy soil, the content of MgO also differs clearly. The content of non-basic elements such as phosphorus is also highest in weakly-leached soils and lowest in strongly-leached soils.

The distribution of the three types of paddy soil follows a zonal regularity. Generally speaking, paddy soils in North China are chiefly of the weakly-leached type, those in Central China are mainly of the moderately-leached type, and those in South China are mostly of the strongly-leached type. However, exceptions are not uncommon. For instance, in the Sunghua River plain and Liaoho plain of Northeast China, there are paddy soils of the moderately-leached type, while in the lower reaches of major rivers in South China there are large areas of paddy soils of the moderately-leached type. Besides, in South China there are also some weakly-leached paddy soils which derive their properties from the parent soil.

REFERENCES

(1) Institute of Soil Science, 1961. Soil Environment of High-yield Rice. Chapters 2,4,9. Science Press, Beijing.
(2) Institute of Soil Science, 1978. Soils of China. Part B, Chapter 2. Science Press, Beijing.
(3) Thorp, J., 1941. Geography of the Soils of China. Geological Survey of China.
(4) Ling Yun-xiao and Yu Tian-ren, 1957. Soil acidity in relation to exchangeable hydrogen and aluminum. Acta Pedologica, **9**: 234–245.
(5) Ding Chang-pu and Yu Tian-ren, 1958. Studies on oxidation-reduction processes in paddy soils. IV. Activities of iron and manganese in paddy soils derived from red soils. Acta Pedologica, **6**: 99–107.
(6) Yu Tian-ren and Ding Chang-pu, 1958. On the status of exchangeable bases and its relation to the genesis of paddy soils derived from red soils. Soils Bulletin, **33**: 31–43.
(7) Yu Tian-ren, et al., 1959. Formation and amelioration of the low-yield "White soil" of the Taihu region. Acta Pedologica, **7**: 42–58.
(8) Zhang Xiao-nian, 1961. Clay minerals of paddy soils of China. Acta Pedologica, **9**: 81–102.
(9) Cao Sheng-geng and Yao Yu-cheng, 1964. Pedogenetic horizons of paddy soils and their classification. Soils Bulletin, **36**: 179–205.
(10) Zhang Xiao-nian, 1981. Changes in clay minerals in the genesis of paddy soil. in "Proc. Symp. Paddy Soil". pp. 475–479. Science Press, Beijing. (in English)

CHAPTER 10

SOIL AND PLANTS

Yu Tian-ren

Most of the physico–chemical properties of paddy soil as well as the physiological characteristics of the rice plant are related to submergence. In this chapter, in discussing the interrelationship between paddy soil and the plant, special attention is paid to this concept. In the discussions some experimental results relating to upland plants are also cited for comparison. The most important physico–chemical properties of the soil in relation to plants are oxidation–reduction status, pH, and nutrient status, which will be dealt with in this chapter. Reviews regarding this problem (not including Chinese literature) have been written by Tadano and Yoshida[17] and by Ponnamperuma[18].

10.1 OXIDATION–REDUCTION STATUS

10.1.1 Adaptation of rice plants to oxidation-reduction conditions of the soil

Physiologically rice is distinguished from upland crops in that it can use the oxygen of the atmosphere by absorption through leaves, and is also capable of transferring the oxygen from leaves to roots. Therefore, if there were not the involvement of other chemical factors, the low oxidation–reduction potential per se would have no unfavorable effect on the growth of rice within a certain *Eh* range. This viewpoint can be illustrated by a comparison between the behavior of rice and lettuce shown in Table 10–1. Lettuce grown on a yellow-brown soil which had been submerged for several months showed symptoms of white spots on the leaves and gave a yield of only 1/4 of that grown under normal conditions due to the preservation of a reduced condition of the soil, although the water content of the soil was kept at 60% of the water–holding capacity during the growing period of the lettuce. On the other hand, the dry weight of rice grown on the reduced soil was even greater than that grown on the original soil. Under field conditions rice can grow normally even after a temporary drying of the soil, indicating that the adaptation of rice to oxidation condition of the soil is also strong.

10.1.2 Influence of rice root on the oxidation-reduction status of the soil

The adaptation ability of rice plants to strong reducing conditions of the soil is also related to the fact that the rice-root can secrete the oxygen coming from

TABLE 10–1 Effect of oxidation–reduction condition of the soil on plant growth[4]

Treatment	Dry weight (g/pot)	
	Rice	Lettuce
Submerged for several months	17.8	5.0
Dried after submergence	17.8	21.3
Original soil	10.1	18.4

the leaves to the soil through the rhizospheres. This peculiarity in the physiology of rice has been demonstrated by a series of measurements. For example, it can be seen from Fig. 10–1 that in the cultivated layer of a paddy field where rice roots are abundant the oxygen content in the root–zone is higher than that in the soil without roots. This oxygen would induce a change in a series of chemical properties in the root–zone, in which the increase in oxidation–reduction potential is most notable (Fig. 10–1). This peculiar physiological behavior of the rice root will be more evident if a comparison is made between the Eh of the root–zone (rhizosphere soil) of rice and that of upland crops. It is seen from Table 10–2 that one week after the transplanting of seedlings the Eh_7 of the soil near roots of wheat, soybean and buckwheat was lower than that of the bulk (non–rhizosphere) soil without the influence of plant roots, due apparently to the secretion of organic reducing substances from the roots. On the contrary, the Eh_7 of the root–zone of rice was higher than that of the bulk soil by 55 mV. For plants which have grown to a height of 20 cm this difference was more distinct, as evidenced in Tables 10–3 and 10–4. The Eh of the soil where root system

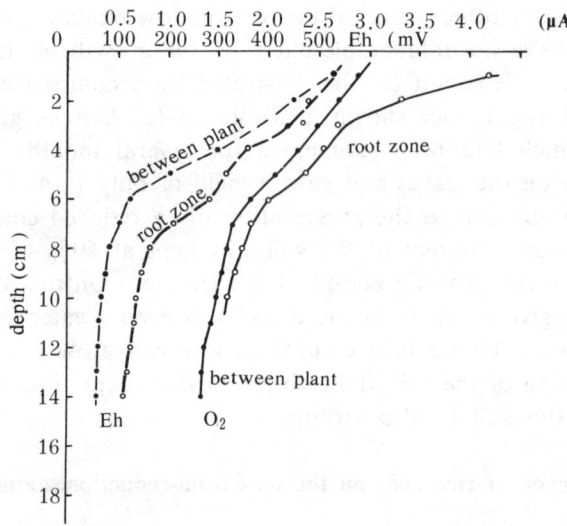

Fig. 10–1 Comparison of O_2 content and Eh between root–zone of rice and bulk soil (Jiangxi)[1]

10.1 Oxidation-reduction status

of wheat was dense was lower by 70—190 mV than that of the bulk soil without roots, whereas the Eh in the root-zone of rice may be higher by 250—350 mV. It should also be noticed that the difference of Eh in the horizontal direction as well as in the vertical direction was very regular, indicating that the denser the roots in the soil, the larger the influence on the Eh of the soil.

TABLE 10-2 Influence of plant roots on Eh_7 of the soil[13]

Plant	Eh_7 (mV)						Difference between extremities
	1 mm*	2 mm	3 mm	4 mm	5 mm	10 mm	
Wheat	440	458	467	488	495	514	−74
Soybean	554	561	566	567	567	568	−14
Buckwheat	238	247	276	281	278	277	−39
Rice	276	264	259	241	226	221	+55

* Distance from roots

TABLE 10-3 Influence of wheat root on Eh of the soil[4]

Depth (cm)	Eh (mV)		
	0 mm*	10 mm	20 mm
0—3	425	485	605
3—8	545	535	615
8—15	615	610	610
15—20	620	620	610
20—30	n.d.	625	605

* Distance from roots

TABLE 10-4 Influence of rice root on Eh of the soil[4]

Depth (cm)	Eh (mV)			
	0 mm*	5 mm	10 mm	20 mm
0—3	250	80	15	−30
3—5	150	30	0	−60
5—8	155	5	−15	−80
8—15	70	−15	−55	−110
15—20	55	−30	−65	−95
20—30	30	−45	−80	−110

* Distance from rice roots

The oxidation–reduction potential is an overall reflection of the kind and the amount of the oxidation–reduction substances of the soil. The micro–regional difference in Eh is in essence a reflection of the difference in reducing substances of the soil. It has been observed that at the booting stage of rice growing on a paddy soil derived from red soil, the amount of reducing substances in the root–zone is only 37% of that in the bulk soil. Hence, under different planting densities the amounts of various reducing substances in the soil differ distinctly, as shown in Table 10–5. The Eh of the cultivated layer increased with the increase in planting density, and correspondingly, the amounts of ferrous iron, manganous manganese and organic reducing substances decreased.

TABLE 10–5 Oxidation–reduction regime of the soil as affected by planting density of rice (Zixi)[12]

Item	Planting density (kilo-buches/ha)	Growing stage				
		Tillering	Elongating	Booting	Earing	Milky
Manganous manganese (m.e./100g)	300	0.10	0.12	0.05	0.12	0.02
	600	0.10	0.11	0.03	0.08	0.01
	900	0.09	0.09	0.02	0.03	0.01
Ferrous iron (m.e./100g)	300	5.34	8.49	5.14	8.16	0.98
	600	5.24	7.53	4.18	4.61	0.74
	900	4.63	5.95	1.79	1.83	0.61
Active reducing substances (m.e./100g)	300	5.71	9.15	5.89	9.73	1.37
	600	5.41	8.08	4.59	5.16	1.13
	900	4.83	6.32	3.85	4.35	0.86
Total reducing substances (m.e./100g)	300	7.81	10.9	6.21	10.3	1.75
	600	7.66	9.54	5.08	7.32	1.60
	900	7.08	8.82	4.52	4.56	1.35
Eh (mV)	300	105	45	100	30	250
	600	115	80	135	120	290
	900	145	130	185	160	340

All of the data cited above lead us to believe that the influence of rice roots on the oxidation–reduction status of the soil is considerable. This should be the basic reason why rice can grow normally in reduced soils.

10.1.3 Influence of the oxidation-reduction condition of the soil on the Eh of plants

Considering the question from another viewpoint, the oxidation-reduction condition of the soil would exert influences on the oxidation-reduction status of plants. This is true for both upland crops which are less tolerant and rice plants which are more tolerant to the reducing conditions of the soil. This will be illustrated in the following discussion, taking Eh as an index.

Lettuce seedlings were transplanted in soils with different oxidation-reduction conditions, and after one week the Eh of the plant was measured by inserting a platinum electrode into the stem. The mean values of 11 determinations were: 635 mV when grown in the soil with an Eh of 590 mV, 618 mV when grown in the soil with an Eh of 400 mV, and 603 mV when grown in the soil with an Eh of —200 mV [4]. Fig. 10-2 shows the depolarization curves of the platinum electrode inserted in lettuce growing in two soils. It can be seen from the anodic depolarization curve that the difference in final potential readings was 50 mV. For the cathodic depolarization curve it was 75 mV. In spite of the fact that the difference in the Eh of the plants was not so large as the soil (655—515=140 mV), the data does show that lettuce grown on an oxidizing soil has a stronger oxidation system.

The influence of soil conditions on rice plants in this respect is also quite clear. In an experiment, the oxidation-reduction condition of the soil was regulated by adding easily decomposable organic matter or hydrogen peroxide, and the Eh of the expressed leaf sap of rice was measured periodically. It can be seen from the data shown in Fig. 10-3 that the difference in the Eh of plant sap was parallel to the difference in the Eh of the soil : rice growing on a reducing soil had a lower Eh in the leaf sap.

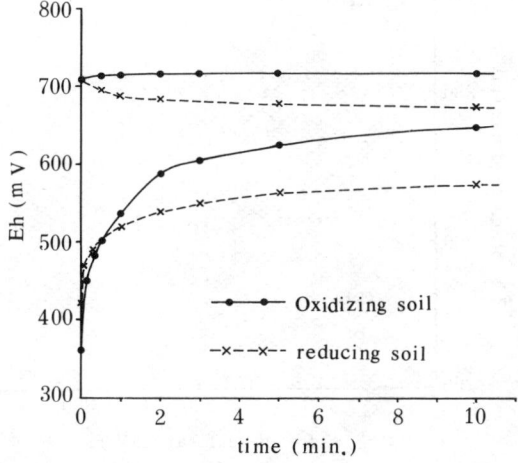

Fig. 10-2 Depolarization curves of Pt electrode in lettuce grown on soils of different oxidation-reduction status[4]

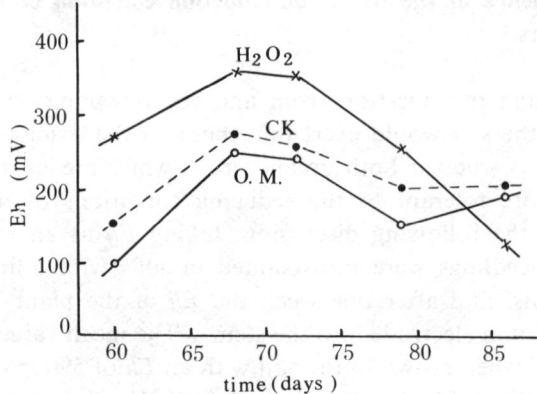

Fig. 10-3 Influence of oxidation-reduction condition of the soil on the *Eh* of leaf sap of rice[5]

It may be expected that if a quinhydrone solution is added to leaf sap the oxidation-reduction potential of quinhydrone will change due to the change in the ratio of quinone (oxidant) to hydroquinone (reductant) caused by a chemical reaction with the oxidation-reduction substances of the sap, and that the magnitude of *Eh* change will be dependent on the kind and the amount of the oxidation-reduction systems of the leaf sap. Actually, there exists a general tendency to a decrease in the *Eh* of the quinhydrone solution with a decrease in the *Eh* of the leaf sap added to the solution, as is seen in Fig. 10-4. The scattering of points

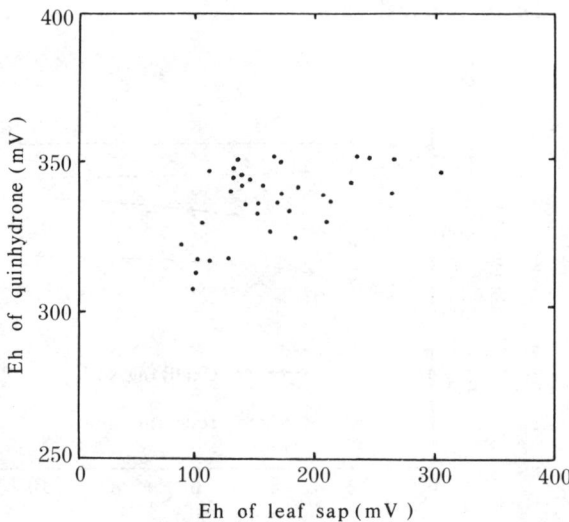

Fig. 10-4 Relationship between *Eh* of quinhydrone and *Eh* of the leaf sap of rice[5]

10.1 Oxidation-reduction status

from a definite line is assumed to be partly due to the variation in pH of leaf saps, which is conducive to the change in quinone to hydroquinone ratio through affecting the dissociation of the latter. As this change in the *Eh* of the quinhydrone solution also reflects the difference in the amount of oxidation-reduction substances of leaf saps, it must be considered from the data of Figs. 10-3 and 10-4 that the oxidation-reduction condition of the soil can affect the composition of oxidation-reduction substances in rice plants.

So far, the physiological significance of the difference in oxidation-reduction potential of plants is not known.

10.1.4 Toxicity problems under strong reducing conditions

Although rice plants can tolerate the reducing conditions of the soil to a certain extent, there is evidence that a strong reducing condition is toxic to rice growth. At present a controversy about the principal causes of the toxicity under different environmental conditions still exists among scientists. In the following, on the basis of experimental results obtained in China, the toxicities of ferrous iron and hydrogen sulfide, which seem to be the more important problems, will be discussed.

It is well known to farmers in many districts of China that the growth of rice will be hampered if rusting colloidal materials exist on the surface of standing water in the field, or if rusting waters are used for irrigation. However, this adverse effect will disappear if the rusting materials are removed from the water by the addition of lime before irrigation. Many field measurements show that in strongly reducing soils where rice plants show symptoms of abnormal growth, the content of water-soluble ferrous iron is generally high. It was observed in a pot experiment[5] that there was a significant negative correlation between the growth of rice and the ferrous iron content of the soil. Further experiments showed that the addition of ferrous iron to the soil would result in the retarded growth of rice, and that the higher the amount of ferrous iron added, the more severe the adverse effect (Table 10-6). However, under field conditions the severity of toxicity for rice growth in various soils is not proportional to the content of ferrous iron. This is because there is only a small part of ferrous iron existing in the water-soluble and ionic form, and there may be other factors affecting the growth of the rice. Based on some field data, it is supposed that the critical concentration of water-soluble ferrous iron for rice growth in South China is about 50—100 ppm.

With respect to the problem of hydrogen sulfide, it is necessary to distinguish the total amount of sulfides in the soil from the molecular hydrogen sulfide which is actually responsible for the toxicity to plants. It has been known in Chapter 5 that the amount of hydrogen sulfide in a soil is primarily determined by the precipitation-solution equilibrium between sulfide ions and ferrous iron and manganous manganese, and by the association-dissociation equilibrium between sulfide ions and hydrogen ions. In most paddy soils of China there are sufficient ferrous and manganous ions to react with sulfide ions to form insoluble sulfides. Therefore, the actual amount of hydrogen sulfide in paddy

TABLE 10-6 Toxicity of ferrous iron to rice (pot experiment)[5]

Paddy soil	Fe^{2+} added* (mg/100g)	Fe^{2+} in soil			Height of rice (cm)		Dry weight (g/pot)
		1st day	11th day	27th day	13th day	27th day	
Derived from red soil	0	17.2	32.6	20.6	29.7	50.3	1.89
	50	46.7	59.4	36.4	22.5	41.7	0.68
	100	95.4	86.8	56.6	8.3	10.7	Dead
	200	163.2	162.5	109.2	3.3	9.7	Dead
Derived from granite material	0	4.7	—	9.5	32.5	54.7	2.30
	50	—	24.8	34.1	28.9	51.7	1.87
	100	50.8	46.1	37.5	17.7	27.0	0.27
	200	147.0	154.0	68.9	5.1	11.5	Dead

* Added one week before experiment

soils is generally lower than the critical concentration (about 0.07 ppm) as established by plant physiologists. Thus, the direct toxicity of hydrogen sulfide to rice plants should not be a frequently encountered problem. On the other hand, it was observed that in a limited local area of a newly formed paddy field developed on red soil where rice plants showed symptoms of toxicity, the concentration of hydrogen sulfide could be as high as 0.68 ppm (cf. Chapter 4). Laboratory studies also showed that the concentration of H_2S at pH 4 may have been 1.53 ppm and at pH 5.2, 0.21 ppm[16]. Under such circumstances the possibility of the toxicity of hydrogen sulfide to rice roots should exist. Deduced theoretically, the toxicity of hydrogen sulfide may occur under the following circumstances: (a) Soils containing a high amount of fresh organic matter are submerged suddenly under high temperature. In such cases the amount of ferrous and manganous ions may be insufficient to precipitate all the newly-formed sulfide ions. (b) Strongly-leached sandy soils have a high content of organic matter. In this case the amount of reducible ferric oxides may be too low. (c) Dried acid sulfate soils are planted for rice immediately after submerging. In such cases there is the possibility of the presence of a high concentration of molecular hydrogen sulfide due to the low pH of the medium.

The toxicity of ferrous iron and sulfide on the growth of rice is manifested clearly in the formation of black roots. Black roots may be distinguished as two kinds. One is a "chemical" black root, the surface of which is covered with black-colored sulfides. The sulfur content in black roots as determined with S^{35} was higher by 3—4 times than that in yellow-red roots[3]. It can be seen from Fig. 10-5 that the proportion of black roots is closely related to the oxidation-reduction status of the soil, i.e., the lower the Eh, the higher the percentage of black roots. Drying the field can induce a decrease in the percentage of black roots from 36.3% to 8.6%[1]. The black color of the root will fade after exposure to air for 1—2 hours. This kind of root is only slightly affected and the physiological function can be restored if the oxidation–reduction condition of the soil is improved. Another kind of black root is physiological, usually making its appearance when rice is grown on strongly reducing soils. It cannot be turned to a yellow–red root even after the drying of the field.

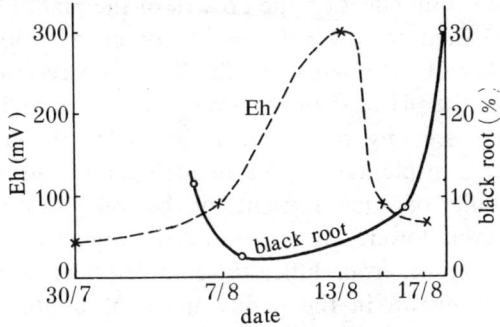

Fig. 10–5 Relationship between black roots of rice and Eh of the soil ("white soil")[1]

10.2 pH

Soil pH can affect the growth of plants indirectly through affecting a series of other properties of the soil. In the following, only some direct aspects will be discussed.

It is generally considered that rice plants can adapt to a wide range of pH of the medium. A pot experiment conducted with a paddy soil derived from red soil showed that within the pH range of 5—7 the effect of pH on three varieties of rice was insignificant (Fig. 10–6). As known in Chapter 7, the pH

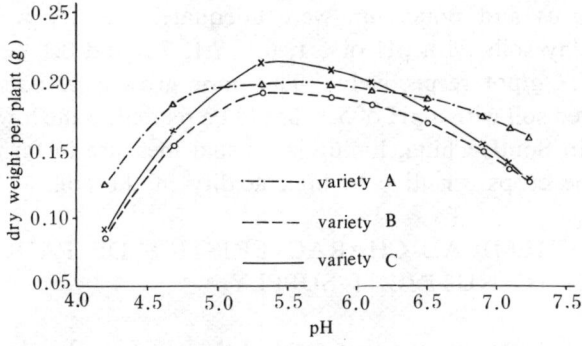

Fig. 10–6 Effect of soil pH on rice growth (courtesy of J. C. Xie)

of most paddy soils under submerged conditions is within the range of 6—7.5. However, the pH of some alkaline soils may increase to higher than 9 when submerged due to the intensification of the hydrolysis of some salts. In such cases an unfavorable high pH effect on rice roots is possible. It should also be remembered that pH is the overall reflection of some chemical properties of a soil, and it will in turn affect a series of other properties of the soil. Therefore, under natural conditions the harmful effect on the growth of the plant of some chemical factors accompanied by high or low pH should not be overlooked.

The high adaptability of rice plants to soil pH is to a certain degree related to their ability to change the pH of the root–zone. The data in Table 10–7 show that the pH in the rice seedling root–zone is lower by 0.2 unit than in the bulk soil one week after transplantation to a neutral paddy soil, which would be beneficial to the availability of some nutrients of the soil. The pH of the root–zone of buckwheat is even lower. It is known that buckwheat has a strong ability to obtain its phosphorus from difficultly–soluble phosphates. Some green manure plants commonly grown in the paddy fields of South China, such as radish and milk vetch, also have such an ability.

TABLE 10–7 Influence of plant root on pH of a neutral paddy soil[13]

Plant	pH						
	1 mm*	2 mm	3 mm	4 mm	5 mm	10 mm	Difference between extremities
Buckwheat	5.10	5.30	5.88	6.11	6.21	6.22	−1.12
Wheat	6.93	6.95	6.94	6.94	6.94	6.95	−0.02
Rice	6.75	6.78	6.82	6.86	6.92	6.95	−0.20

* Distance from roots

On the other hand, for some upland crops commonly planted on paddy soils, low pH is frequently an important factor responsible for a low yield. For example, it was found in a pot experiment that under conditions where the supplies of nitrogen, phosphorus and potassium were adequate, the dry weights of oats grown on five paddy soils with pH of 5.1, 6.1, 7.1, 7.2 and 8.1 were 21.5, 43.2, 33.6, 33.0 and 26.8 g/pot respectively. The poor growth of oats on a paddy soil derived from red soil with a pH of 5.1 should be related to the low pH. Therefore, for farmers in South China, liming is a usual measure in farming practice, especially for some crops sensitive to high acidity in the soil.

10.3 PHYSICO–CHEMICAL CHARACTERISTICS OF PADDY SOIL IN RELATION TO NUTRIENT SUPPLY

It has already been mentioned that most of the physico–chemical characteristics of paddy soil are related to submergence. The interrelationship between these characteristics and nutrient supply is related to many considerations. In the

10.3 Physico-chemical characteristics of paddy soil

following paragraphs, the form and movement of some nutrients will be discussed at length.

10.3.1 Forms of nutrients

The oxidation–reduction condition is a major factor in influencing the transformation of nitrogen in paddy soil. A reduced condition is beneficial to the accumulation of organic nitrogen[1]. For the mineral part of nitrogen, there is a negative correlation between the ratio of ammonium–nitrogen to nitrate–nitrogen and the oxidation–reduction potential of the soil[14]. Under laboratory conditions the denitrifying intensity in paddy soils is affected by aeration[3]. It is commonly believed that the low utilization rate of mineral nitrogen–fertilizers in paddy soils is due chiefly to the transformation of ammonium–N to nitrate–N under certain oxidizing conditions and the subsequent volatilization loss caused by denitrification under certain reducing conditions. It should also be recalled that the tendency to approach to neutral reaction under submerged conditions is helpful to the progress of these reactions.

The presence of a large amount of ferric oxides is an important factor responsible for the low availability of phosphorus in the paddy soils of South China. On the other hand, the water regime of the soil has a distinct influence on the availability of phosphorus. It is generally observed that the yield of upland crops is extremely low when they are planted on paddy soils derived from red soils without the application of phosphorus fertilizer, whereas the yield of rice may be considerable under similar conditions. For instance, in a pot experiment conducted with a young paddy soil derived from red soil, the yield of oats for the check treatment was only 3—5% of that of the treatment fertilized with phosphate, whereas the yield and the utilization rate of phosphorus for rice were 39% and 30% respectively[8]. Further studies revealed that ferric phosphate in the paddy soils of South China could be mobilized gradually with the prolongation of submergence of the soil, and that the "A" value of phosphate at the maturity stage of rice growth was 2—5 times that of the early–tillering stage[3]. All of these data show that the reduction condition of the soil is beneficial to the availability of phosphorus. Therefore, in agricultural practice it could be expected that to transplant rice seedlings after a certain period of submergence, or to apply a small amount of phosphate fertilizer near the root–zone in the early period of tillering when the concentration of available phosphate of the soil is still low, would alleviate the deficiency of phosphorus to a certain degree at the early period of rice growth.

The effect of the physico–chemical conditions of the soil on the availability of some minor elements is also important. For example, in recent years it was found that in some areas of China the paddy soil was deficient in zinc. What is interesting is that this occurs only in calcareous soils, and that the soil may be adequate in zinc supply when upland crops are planted. As it has been shown in Chapter 5 that the addition of zinc to reduced soils can lead to a decrease in the concentration of water–soluble sulfide, there is reason to assume that this phenomenon of zinc deficiency for rice is related to the precipitation of zinc sulfide under reduced conditions at a high soil pH. The situation with manganese is different. Inasmuch as under submerged conditions most of the manganese in the soil is reduced to the manganous form, generally speaking there should be no problem of manganese deficiency for rice. However, because the amount of exchangeable (including water–soluble) manganese is strongly dependent on the pH of the soil, the possibility of manganese deficiency in some calcareous paddy soils when planted for upland crops should not be excluded. As a matter of fact, the uptake of manganese by rice is dependent on the pH of the soil. It is seen from the results of a pot experiment shown in Table 10–8 that the content of manganese in the stalk of rice decreased regularly with the increase in pH of the soil due to the application of $CaCO_3$, $MgCO_3$ or Na_2CO_3. From this viewpoint, it must also be imagined that toxicity to rice due to excessive amounts of manganese in some strongly acid paddy soils, such as acid sulfate soils in the coastal area of China, should be possible.

TABLE 10–8 Effect of soil pH on manganese content of rice[7]

Treatment		Dried soil		Mn content of rice (mg/100g)
Substance	Amount (g/pot)	pH	Exchangeable Mn (mg/100g)	
CK	0	5.8	20.7	220
$CaCO_3$	10	7.1	5.0	205
	20	7.7	tr.	56
	40	7.9	tr.	42
$MgCO_3$	8.7	6.6	10.6	145
	17.3	7.0	1.3	145
	34.6	8.0	tr.	50
Na_2CO_3	10.6	7.3	7.4	73
	21.2	8.0	tr.	10

10.3.2 Movement of nutrients

The current assumption is that the availability of nutrients for plants is dependent on the amount of the readily available form and the rate of transformation from the potentially available form to the readily available form as well as on the migration rate from the soil to the root surface. There are two mechanisms, i.e., ion diffusion and mass flow, which are responsible for the migration of nutrient ions in the soil. As far as paddy soil is concerned, these mechanisms are all favoured by submergence. This is because under water–saturated condi-

10.3 Physico-chemical characteristics of paddy soil

tions the diffusion coefficients of various ions are much larger than those under upland conditions[2]. Besides, in large areas of paddy soils derived from red soils in South China, the diffusion rates of some nutrient ions are higher than those in their parent soils because of deep changes in chemical composition and mechanical composition during the pedogenetic processes. For example, in an experiment in which the diffusion of soluble components of granulated superphosphate in two soils was compared, taking electric conductance as an index, it was found that after two weeks the diffusion distance in the red soil was about 28 mm, whereas in the corresponding paddy soil, it was about 42 mm (Fig. 10–7).

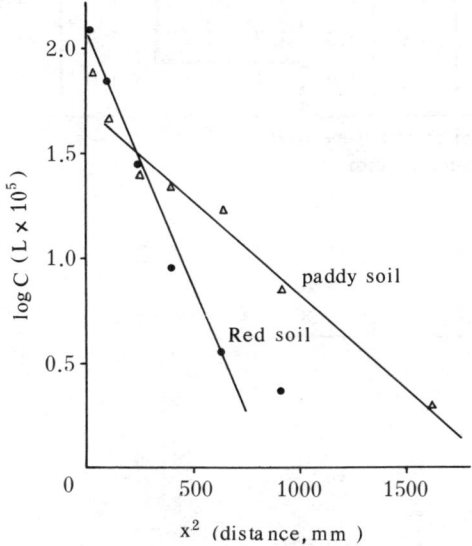

Fig. 10–7 Diffusion of ions from granular superphosphate in soils[13]

The diffusion rate of fertilizers composed of monovalent ions is even higher. Table 10–9 shows the activity (concentration) of ammonium ions in the vicinity of granulated ammonium bicarbonate after being placed in a neutral paddy soil. It is evident that ammonium ions have diffused to a distance of 2 cm on the second day of placement, and to as far as 6—7 cm on the sixth day. On the 19th day the ammonium ions remaining at a distance of 1 cm decreased to a very low level, due apparently to the diffusion to a further distance.

Mass flow of nutrient ions in paddy soils is also an important mechanism, and in many cases it is more important than diffusion. Under submerged conditions soil water percolates at a certain rate, and this downward movement of nutrients along with gravitional water forms another driving force added to the concentration gradient responsible for the migration of nutrients. Fig. 10–8 shows an example in this respect. After the placement of a granule made of ammonium bicarbonate at a depth of 7 cm in a neutral paddy soil, the distribution in electrical conductivity at a distance of 1 cm horizontal to the granule was

TABLE 10–9 Change in NH_4^+ concentration near granular ammonium bicarbonate[15]*

Time (days)	NH_4^+ (N×10³)									
	1**	2	3	4	5	6	7	8	9	10
0.5	22	1.0	1.0	1.0	1.0	1.0	1.1	1.0	—	—
1	29	2.3	1.1	1.1	1.0	1.1	1.1	1.0	1.1	—
3	33	12	1.5	1.3	1.1	1.1	1.0	0.9	1.2	1.1
5	36	23	4.2	2.0	1.7	1.5	1.3	1.2	1.2	1.1
7	23	15	10	3.5	1.3	1.2	1.0	1.0	1.0	0.9
11	12	8.8	5.6	2.5	1.2	1.2	1.0	1.0	1.0	0.9
18	5.1	4.6	3.1	2.9	1.1	1.1	1.0	0.9	1.0	1.0

* Determined with glass micro-electrode
** Distance from granule in mm

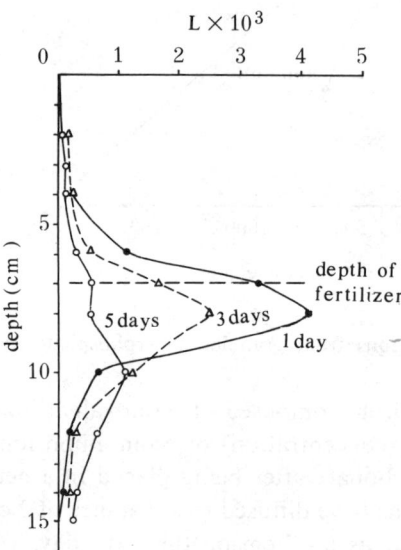

Fig. 10–8 Vertical migration of ammonium bicarbonate in paddy soil (determined at 1 cm beside the granule)[15]

not symmetrical. The concentration peak was at a depth below that of the granule. And, the longer the duration of experiment, the more distinct the asymmetry. Apparently this is due to the downward movement of ions together with percolating water.

In addition to the mechanism mentioned above, nutrient ions can also migrate toward the root surface along with the water absorbed by plant roots.

10.3 Physico-chemical characteristics of paddy soil

In spite of the high rate of nutrient movement by ion diffusion and mass flow in paddy soils (this is also an important reason why rice is a high-yield crop), in field conditions the absorption rate of nutrients by rice is generally higher than the migration rate toward rice roots. As a result, the concentration of nutrient ions is generally lower in the root-zone than in the bulk soil. This can be illustrated by the following examples.

Consider the total concentration of ions. It can be seen from Fig. 10-9 that the nearer the soil is to the root system of the rice, the lower is its electrical conductivity. The difference between two extremities may amount to 37%. The higher electrical conductivity near the roots of buckwheat is supposed to be due to the higher concentration of H^+ ions with high mobility and the dissolution of some elements at a low pH. Under field conditions it was observed that in the tillering-elongating period the electrical conductivity in the root-zone of rice was always lower than that in the bulk soil (Table 10-10). This difference in conductivity is apparently due to the movement of nutrient ions toward the root-zone lagging behind the absorption by rice roots.

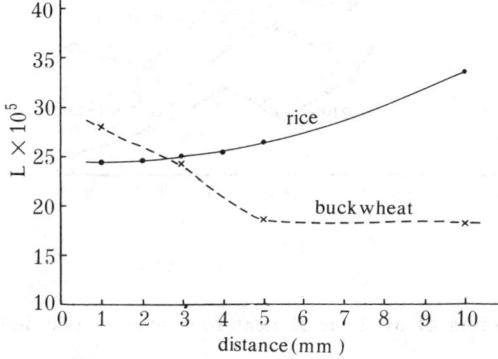

Fig. 10-9 Electrical conductivity near root-zone of rice and buckwheat[13]

TABLE 10-10 Dynamics of electric conductance in root-zone of rice and bulk soil[10]

Treatment	Location	Electric conductance ($L \times 10^5$)								Mean
		7/8*	13/8	16/8	22/8	27/8**	9/2	7/9	16/9	
Lightly fertilized	Root-zone (A)	27.6	43.6	52.9	38.6	30.8	50.6	38.2	23.3	38.2
	Bulk soil (B)	29.7	51.2	50.0	46.6	37.6	60.9	47.0	25.2	43.5
	(B/A)×100	108	118	95	121	122	120	123	108	114
Heavily fertilized	Root-zone (A)	35.7	48.4	51.2	48.0	34.6	48.5	40.2	26.4	40.4
	Bulk soil (B)	38.9	57.3	61.0	60.2	35.4	66.3	43.9	23.2	48.3
	(B/A)×100	109	118	119	125	102	137	109	142	120

* Date
** Topdressing with N at Aug. 29

If individual species of nutrient ions are considered, the difference between the root–zone and the bulk soil is more distinct than that of the total ion concentration. Figs. 10–10 and 10–11 show measuring results for available potassium and ammonium–nitrogen. It can be seen that the amounts of nutrients in the root–zone were much lower than in the bulk soil during the whole growing period of rice in both the lowly– and highly–fertilized fields.

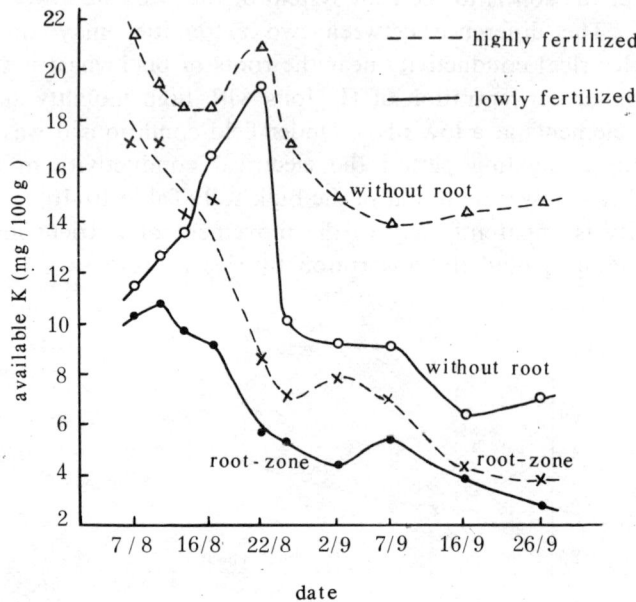

Fig. 10–10 Dynamics of available K near root–zone of rice and bulk soil[1]

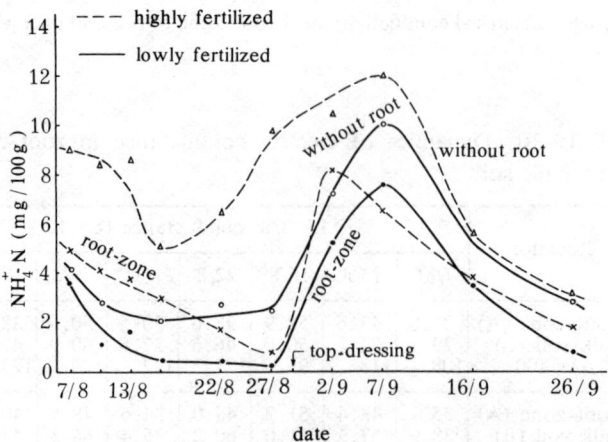

Fig. 10–11 Dynamics of NH_4^+–N near root–zone of rice and bulk soil[1]

10.3 Physico-chemical characteristics of paddy soil

The percentage of monovalent ions such as ammonium and potassium in the liquid phase is higher than that of polyvalent ions such as calcium and phosphate. When the monovalent ions migrate in the soil, the retarding action exerted on them by the electric charge of clay particles would be weaker than that on the polyvalent ions. Therefore, there are plenty of reasons to suspect that the difference in the amount of available phosphate between the root–zone and the bulk soil would be more distinct than that mentioned in the last paragraphs.

A question pertinent to the proper method of fertilization arises here. Theoretically, it should be advantageous to apply both phosphate fertilizers and nitrogen and potassium fertilizers in a position close to the roots of rice (or other crops). That this is true has been verified by the long–term experience of farmers of China and by some experimental results obtained in recent years. We shall also say a few words about a new theory prevalent in the last two or three decades. According to this theory, in order to reduce the fixation of phosphate by soils, especially by those soils containing large amounts of iron and aluminum oxides, it is advantageous to apply phosphate fertilizers in a granular form. However, it must be remembered that when used in a granular form the chance of the fertilizer contacting plant roots will be less than when it is used as a powder due to the reduction in outer surface area. Hence the availability for plants will also be lowered, particularly in the case of phosphates with a low migration rate. From an experimental result shown in Fig. 10-12 it can be seen that even for a paddy soil containing a large amount of ferric oxides the granular superphosphate did not show a superiority to powdered superphosphate, provided that they were all placed close to the rice roots. In the former case the utilization rate of applied P^{32} was 14.5% and the proportion of phosphorus coming from fertilizer was 18.4% of the total phosphorus in the rice plant, whereas in the latter case the corresponding figures were 15.4 and 19.0% respectively. Of course, when the superphosphate was mixed with the whole soil of the cultivated layer in powdered form, the corresponding figures were only 5.0% and 5.4% respectively. These results demonstrate the importance of placing fertilizers concentratively in the vicinity of

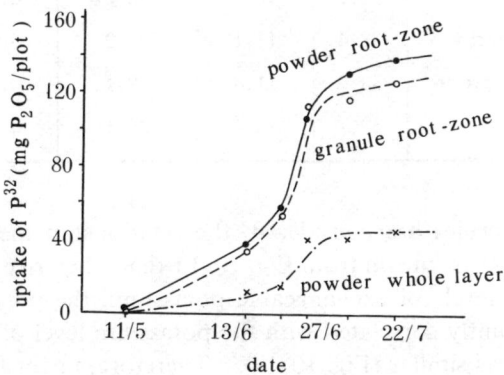

Fig. 10–12 Effect of method of application on utilization of P–fertilizer[11]

plant roots, and also indicate that the primary prerequisite for fertilization is to promote contact between fertilizer and plant roots as much as possible. Considered from this viewpoint, it may be assumed that a proper permeability in paddy soil is beneficial to the absorption of nutrients by rice.

10.4 NUTRIENT CONTENT OF THE SOIL IN RELATION TO UPTAKE BY RICE

According to our present knowledge about the mechanism of nutrient uptake, the surface of plant roots possesses an ability for ion-exchange, and the initial stage of nutrient uptake is an ion-exchange reaction between the root surface and the soil solution. Hence, in addition to the selective ability of the plant, the uptake of nutrients is profoundly affected by the ionic composition of the soil solution. This is also the case with rice. Table 10-11 shows the relationship between the total ion concentration for rice plants and that of the soil. In this table, the ion concentration in the rice as determined by electrical conductivity measurement with a pair of micro-electrodes inserted into the root or petiole of the rice plant increased regularly with the increase in ion concentration of the soil. As a result, the ratio of the specific conductance in the rice roots to that in the soil remained for the five plots at a constant of 0.31 ± 0.02. The ion concentration in functional leaves increased more distinctly with the increase of ion concentration in the soil, resulting in a conductance ratio of 0.45 for the poorest soil and 0.76 for the most fertile soil.

TABLE 10-11 Relationship in electrical conductance between rice and soil[13]

Plot no.	Appearance of rice	Electrical conductance ($L \times 10^5$)			Ratio	
		Soil	Root	Functional leaf	Root/Soil	Leaf/Soil
1	Excessively yellow	26.8	8.1	12.1	0.30	0.45
2	Normal yellow	30.7	10.2	13.6	0.33	0.44
3	Normal green	38.4	11.1	24.2	0.29	0.63
4	Excessively green	38.9	11.4	29.1	0.29	0.75
5	Weeping	39.7	12.6	30.1	0.32	0.76

If individual ion species are considered, the relationship mentioned above can also be observed. It is known from Fig. 10-13 that when rice was planted in five soils with different levels of exchangeable potassium, the potassium content in the stalk was significantly correlated with the potassium level of the soil. For calcium, the tendency was similar (Fig. 10-14). Therefore, under field conditions it may be frequently observed that the contents of nutrients for the same variety of rice and at the same growing stage vary greatly, depending on the nutrient

10.4 Nutrient content of the soil in relation to uptake by rice

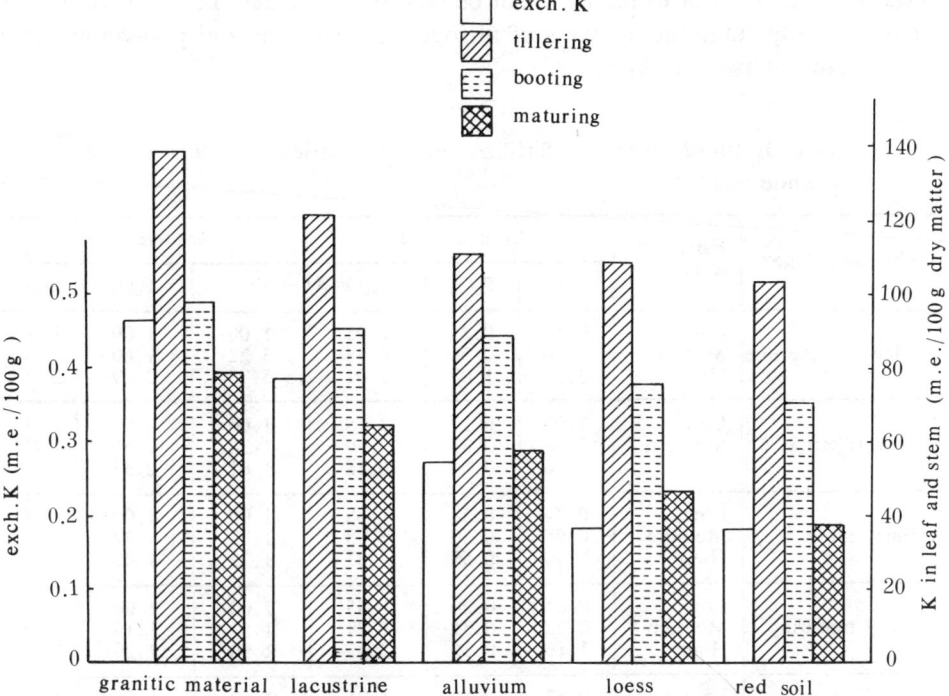

Fig. 10–13 Relationship between K content of rice and exchangeable K of the soil[6]

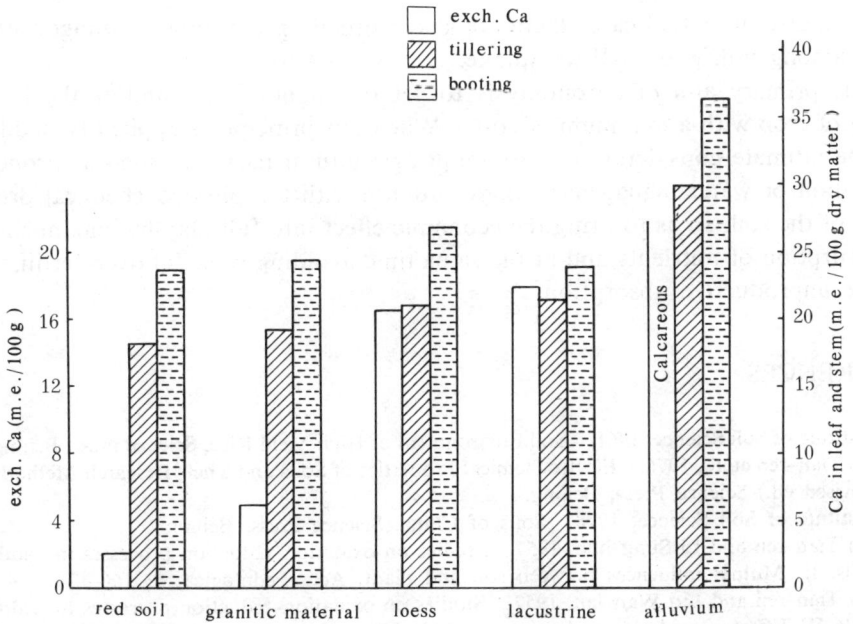

Fig. 10–14 Relationship between Ca content of rice and exchangeable Ca of the soil[6]

level of the soil. For example, it can be seen from the data of Table 10–12 that at the tillering stage the contents of nitrogen, phosphorus and potassium varied by a factor of two or three.

TABLE 10–12 Effect of fertilizer level on nutrient contents of rice ("white soil")[9]

Growing stage	Fertilizer level	Content (%)			Relative value		
		N	P_2O_5	K_2O	N	P_2O_5	K_2O
Vigorous tillering	Low	0.59	0.35	2.93	1.00	1.00	1.00
	Medium	1.90	0.56	4.38	3.22	1.60	1.50
	High	3.39	1.08	7.01	5.75	3.09	2.39
Late tillering	Low	1.03	0.27	2.15	1.00	1.00	1.00
	Medium	1.38	0.42	3.60	1.34	1.56	1.67
	High	1.66	0.64	6.28	1.61	2.37	2.92
Early milky	Low	0.74	0.38	1.68	1.00	1.00	1.00
	Medium	0.98	0.50	2.08	1.32	1.32	1.24
	High	0.90	0.53	2.87	1.22	1.40	1.71
Early maturing	Low	0.61	0.26	1.55	1.00	1.00	1.00
	Medium	0.92	0.31	1.88	1.51	1.19	1.21
	High	1.16	0.36	2.35	1.90	1.39	1.52
Maturing	Low	0.42	0.15	1.81	1.00	1.00	1.00
	High	0.97	0.23	1.75	2.31	1.53	0.97

Of course, in actual cases there are also mutually promoting or antagonistic effects among nutrients in their uptake.

The primary aim of agronomy is to get the highest yield and/or the best quality of crop with a minimum of cost. When this principle is applied to paddy soil, the ultimate consideration is to adopt agricultural measures such as proper fertilization or water management in accordance with the physico–chemical properties of the soil, so as to bring the economic effect into full play by maximizing the absorption of nutrients and at the same time avoiding wasteful over–fertilization or unproductive absorption.

REFERENCES

(1) Institute of Soil Science, 1961. Soil Environment of High–yield Rice. Science press, Beijing.
(2) Yu Tian–ren et al., 1976. Electrochemical Properties of Soils and Their Research Methods. (revised ed.) Science Press, Beijing.
(3) Institute of Soil Science, 1978. Soils of China. Science Press, Beijing.
(4) Yu Tian–ren and Li Sung–hua, 1957. Studies on oxidation–reduction processes in paddy soils. II. Mutual influences between soil and plant. Acta Pedologica, 5: 166–174.
(5) Yu Tian–ren and Liu Wan–lan, 1957. Studies on oxidation–reduction processes in paddy soils. III. Effect of oxidation–reduction condition of the soil on rice growth. Acta Pedologica, 5: 292–304.
(6) Yu Tian–ren et al., 1958. Relationship between soil condition and cationic balance in plants. Soils Bulletin, 33: 1–15.

(7) Yu Tian-ren, Ling Yun-xiao, Mu Run-sheng and Liu Wan-lan, 1958. Effect of soil acidity on the activity of manganese. Soils Bulletin, **33**: 16–30.
(8) Liu Wan-lan et al., 1958. Availability of superphosphate in paddy soils as affected by method of application. Soils Bulletin, **33**: 83–90.
(9) Yu Tian-ren et al., 1959. Formation and melioration of the low-yield "White soil" in Taihu region. Acta Pedologica, **7**: 42–58.
(10) Yu Tian-ren et al., 1959. Studies on electrochemical properties of soils. I. Electrical conductivity of paddy soils in relation to soil fertility. Acta Pedologica, **7**: 145–158.
(11) Wu Jun, 1961. On method of economic use of phosphate fertilizers in paddy soil. Soils, **9**: 58–59.
(12) Liu Zhi-guang and Yu Tian-ren, 1962. Studies on oxidation–reduction processes in paddy soils. V. Determination of reducing substances. Acta Pedologica, **10**: 13–28.
(13) Liu Zhi-guang and Yu Tian-ren, 1963. Studies on electrochemical properties of soils. II. Application of micro-electrode in soil research. Acta Pedologica, **11**: 160–170.
(14) Yu Tian-ren and Liu Zhi-guang, 1964. Oxidation–reduction processes in paddy soils and their relations to the growth of rice. Acta Pedologica, **12**: 380–389.
(15) Ji Kuo-liang and Wang Jing-hua, 1978. Liberation and diffusion of granulated ammonium bicarbonate in paddy soil as studied with micro-electrodes. Acta Pedologica, **11**: 182–186.
(16) Pan Shu-zheng, Liu Zhi-guang and Yu Tian-ren, 1982. Chemical equilibria of sulfides in submerged soils as studied with a H_2S-sensor. Soil Sci., **134**: 171–176.
(17) Tadano, T. and Yoshida, S., 1978. Chemical changes in submerged soils and their effect on rice growth. in "Soils and Rice", pp. 399–420. IRRI.
(18) Ponnamperuma, F. N., 1978. Electrochemical changes in submerged soils and the growth of rice. in "Soils and Rice", pp. 421–444. IRRI.